THE SECRET LIFE OF ANIMALS

THE SECRET LIFE OF ANIMALS

Reader's
Digest

PUBLISHED BY

THE READER'S DIGEST ASSOCIATION LIMITED

LONDON NEW YORK MONTREAL SYDNEY CAPE TOWN

THE SECRET LIFE OF ANIMALS
Edited and designed by Toucan Books Limited
with Bradbury and Williams
Written by John Man
Edited by Helen Douglas-Cooper
Picture Research by Marian Pullen

FOR THE READER'S DIGEST UK
Series Editor: Christine Noble
Editorial Assistant: Chloe Garrow
Editorial Director: Cortina Butler
Art Director: Nick Clark

READER'S DIGEST, US
Senior Editor: Fred Dubose
Senior Designer: Judith Carmel
Group Editorial Director, Nature: Wayne Kalyn
Vice President, Editor-in-Chief: Christopher Cavanaugh
Art Director: Joan Mazzeo

Separations: David Bruce Graphics Limited, London

Printed in the United States of America, 1999

Library of Congress Cataloging in Publication Data

The secret life of animals.
    p. cm. -- (Earth, its wonders, its secrets)
  Includes index.
  ISBN 0-7621-0112-1 (hardcover)
  1. Animal behavior. I. Reader's Digest Association. II. Series.
  QL751 .S435 1999
  591.5 – dc21
                                98-7852

FRONT COVER *A few of the 20 000 eggs deposited by the female
lumpsucker fish. Inset: Emperor penguin parents and chicks survive in
the most inhospitable place on Earth, the Antarctic.*

PAGE 3 *The plumage of flamingos gains its distinctive pink tinge from
pigments in the algae they sift from the water.*

CONTENTS

WELL-KEPT SECRETS   *6*

1 PROGRAMMED FOR SURVIVAL   *13*

THE DEBUT   *14*

DEPENDABLE PARENTS   *24*

LEAVING OR STAYING AT HOME   *34*

2 A PLACE TO LIVE   *41*

FINDING A NEW HOME   *42*

LIVING TOGETHER   *50*

HOUSING MATTERS   *60*

3 FIT AND ACTIVE   *71*

HEALTHY BODIES   *72*

OXYGEN, WATER AND TEMPERATURE   *78*

GETTING ABOUT   *88*

4 EATING OUT   *97*

FINDING FOOD   *98*

FEASTING AND FEEDING   *108*

AVOIDING BEING CAUGHT   *118*

5 CONTINUING THE LINE   *131*

RIVALS AND ALLIES   *132*

COURTSHIP AND MATING   *140*

EGGS AND EMBRYOS   *148*

INDEX   *156*

PICTURE CREDITS   *160*

WELL-KEPT SECRETS

The ancient need on the part of humans to watch and study animals in order to catch a meal has, over thousands of years, evolved into a sophisticated scientific study that is steadily revealing the animal kingdom's innermost secrets.

In an age when humans can split the atom, travel to the moon and surf the Internet, it is sobering to realise that our understanding of the natural world is still quite rudimentary. No one, for instance, has yet witnessed an event as large as a great whale being born or explained how a tiny bird finds its way across the world. Nevertheless, inquisitive scientists are now beginning to uncover some of the secrets that have been kept from the human gaze for so long. They have sat patiently recording every subtle movement that an animal makes. They have journeyed to remote and inhospitable places and discovered and described many animals quite new to science. They have entered the extraordinary world of the infinitesimally small to reveal the very workings of nature itself. Their revelations are remarkable, and sometimes unexpected, but they follow a tradition that was established when our ape-like forebears came down from the trees.

Our distant ancestors must have had some awareness of the hidden workings of animals' lives because their very survival depended on a rudimentary knowledge of other species. They would have had to ascertain which were edible, for example, and where those animals could be found, and to understand the behaviour of dangerous predators in order to stay alive.

The earliest attempts to document natural history can be traced back some 5000 years to Babylonian and Assyrian depictions of medicinal plants and veterinary medicine. The Vedic literature of India, dating from around 2300 BC, also includes references to the animal world. There were records, for example, of the cuckoo-like behaviour of the koel, which lays its eggs in the nests of other birds. However, it was not until the rise of the ancient Greek civilisation that the study of animal life became more systematic.

A key figure in this development was the philosopher and scientist Aristotle, born in 384 BC in northern Greece, who became one of the first people to explore and record the activities of animals. Aristotle looked at the function of the parts or organs of animals, and the ways in which animals move and reproduce, and he observed similarities in the behaviour of different animals as well as in the way they looked in order to establish relationships between them. He collected an enormous range of facts, and his *Historia animalium*, which drew attention to the similarities and differences between species in an orderly way, later became the basis for animal classification.

Aristotle recorded the way in which whales and dolphins beach themselves – a mystery to this day – and recognised that dolphins, orcas, sperm whales and right whales are not fish but mammals as they suckle their young and breathe with lungs.

ANCIENT PLANT ART *An Egyptian basrelief wall decoration from Karnak (left) depicts a botanical scene. Right: Notes and illustrations from a ancient Arab manuscript show that thousands of years ago people were as fascinated by the wildlife around them as we are today.*

He saw that toothed whales are quite different from baleen whales and that 'the latter has no teeth, but does have hairs that resemble hog bristles'. He drew attention to the sounds they make, adding that: 'The dolphin, too, utters a murmuring sound that is equivalent to a voice.'

ON CLOSER INSPECTION

Early naturalists used simple observation to understand much about the natural world. The field diaries of the English naturalist Charles Darwin, for example, give extraordinarily detailed accounts of the plants and animals that he discovered in remote parts of the world. His field notes on the finches of the Galapagos Islands helped him to establish his theories of natural selection and evolution.

Careful observation in the field is still the bedrock of biological study, although the secrets it reveals might simply be fresh views of the familiar. Until the 1970s it was always thought, for instance, that boxing hares were jousting males, but now the secret is out. Careful observation by an amateur enthusiast watching a field near his home has revealed that males and females box as part of their courtship ritual. But even this observation required the help of advanced technology. Hares are notoriously difficult to approach, and the hare-watcher gained his more refined view of his subject with the help of a powerful telescope, the kind normally used by astronomers.

Today technology can take us to the remotest parts of the Earth – to the deepest areas of the ocean with the new breed of aquanauts in deep-sea submersibles, or to the high canopy of tropical rain forests with the aid of hot-air balloons, enormous cranes and a delicate scaffolding of lightweight aerial walkways. But technological breakthroughs have also enabled us to observe the inner workings of the nucleus of a cell and unravel the secrets of life itself.

The study of DNA – the genetic blueprint of life in the nucleus of every cell in the body of every living animal – has provided a way of identifying each individual of a species and of revealing who is related to whom, through a process known as DNA –

or genetic – fingerprinting. This is possible because every individual, from a human being to an amoeba, has its own unique genetic fingerprint.

The genetic fingerprinting technique involves extracting the double-chained strands of DNA from the nuclei of cells in a tissue sample, splitting them down the middle, and then chemically 'cutting off' short

pieces using a very specific enzyme. The pieces targeted by the enzyme are the sections, known as 'minisatellites', that vary from individual to individual. These pieces

MAD MARCH HARES *Both male and female European hares 'box' during the courtship rituals of early spring.*

EXTRAVAGANT GURGLER
The cock capercaillie's well-kept secret was its deep voice, pitched at frequencies well below those that can be detected by the human ear.

are labelled with a radioactive marker, and exposed to X-ray film. The resulting picture resembles the bar code seen on goods in shops, the lines representing the different proteins associated with the minisatellites in DNA. Thus, each genetic fingerprint is unique to its owner, much like the conventional fingerprint. The technique is useful to the police, for example, as a sample of blood or body tissue found at the scene of a crime can be compared with a sample

taken from a suspect to reveal if he or she was present.

In marine research, genetic fingerprinting has been used to help count the number of whales in the ocean. Using a crossbow to fire a small dart, which is attached to a piece of string, into a whale's back, scientists take a tiny sample of skin or muscle. When the dart is recovered and the tissues analysed, the resulting information gives scientists an infallible way of identifying that particular whale. A refinement of the technique, known as noninvasive DNA analysis, can be used to identify an individual simply from a hair or feather sample, and whale biologists are now carrying out the analysis on skin cells sloughed off the whale's body and collected in its wake. This avoids the need to

fire a dart into the whale, thus reducing disturbance to the animal.

Although each individual's DNA fingerprint is distinctive, it also contains many similarities that can be used to identify mothers and fathers, brothers and sisters, uncles, aunts and cousins – indeed, entire family trees. It prevents a whale from being counted twice, and is one of many techniques in a modern whale-watcher's toolkit, which also includes photography: scratches and scars on a whale's fins and body, or the black-and-white patterns on a humpback's enormous tail, are photographed and checked against a catalogue to identify individual animals.

Other technological advances help us to see what was once unseen and to hear what was once unheard. Microscopes enable us to visit the world of the infinitely small. The most modern electron microscopes allow us to examine things the size of molecules. X-rays reveal the inner structures of organisms and can also be used to identify fossils that are hidden in rocks. Radar can track night-flying bird migrants; low-light-level night-scopes and infrared cameras help us to see in the dark; and sensitive microphones and hydrophones (underwater microphones) relay the faintest whispers at the highest frequencies, such as those of hunting bats or chirping belugas.

USEFUL HEARING AIDS

Many aspects of the lives of animals are closed to us because of the limitations of our own senses. For example, they produce not only audible sounds but also many that consist of very high and very low frequencies that are beyond the range of human hearing. The invention of the sound spectrograph, an instrument that transcribes sounds as graph patterns, has solved all manner of mysteries.

Naturalists, for example, were always curious as to why the cock capercaillie makes such an insubstantial gurgling sound, not at all the masterful noise one would normally associate with a bird the size and stature of a turkey. In the breeding season the bird visits a lek – an arena where male animals compete with each other for mating stations – fans out its tail, raises its head and bursts forth, not with the deep boom that knowledge of related birds would lead us to expect, but with an insignificant series of clicks and pops.

The mystery was eventually solved with the aid of the sound spectrograph. The pattern of squiggles on the paper revealed that the cock capercaillie does, indeed, make superior sounds but that they are produced at lower frequencies – 40 Hz and below – than we can hear, in a part of the sound spectrum known as infrasound. Closer inspection of the capercaillie's anatomy has revealed that the bird has a Helmholtz resonator in its throat. This consists of a tube that is closed at one end, and when the bird blows exhaled air across the mouth of the

tube, it produces a booming note so deep that we cannot hear it.

The capercaillie shares this acoustic trick with another unlikely sound producer, the king cobra. A snake's hiss was once written off by researchers as a dreary phenomenon, indistinguishable from one species to another, but then one inquisitive researcher armed with a sound spectrograph listened again. The hiss of 21 snakes worldwide, from pythons to vipers and milk snakes to rat snakes (venomous snakes found in India and South-east Asia), was recorded and analysed. The first surprise was that there were dominant frequencies in each hiss, the second that these varied from species to species, although the most common frequency was around 7500 Hz. The king cobra and the mangrove rat snake, however, had much lower dominant frequencies, at 600 Hz and 625 Hz respectively. The scientists then took a look inside.

The king cobra, they found,

UNDERSEA LANTERN
The complicated mouth-parts and teeth of this purple sea urchin were dissected and described by Aristotle.

DEATH HISS *The hiss of the spectacled cobra from India contains sound frequencies that distinguish it from other species of snake.*

has between 24 and 30 pockets lining its trachea, or windpipe, and the rat snake between 14 and 19. These pockets, which probably developed originally to help inflate the 'hood' when an individual rears up to strike, act as resonating chambers. Exhaled air blows across the mouth of each pocket, producing a low-frequency sound. King cobras and mangrove rat snakes do more than hiss – they growl.

GETTING THE INSIDE STORY

Taking a look inside a creature by dissection is a crude but often productive way of revealing, quite literally, its inner secrets. Aristotle was not only an indefatigable collector of animals and animal stories but also a great dissector. By examining the sea urchin, a spherical, spiny creature common on Mediterranean shores, he discovered that it gathers its food by means of a feeding apparatus consisting of 40 skeletal plates bound together by muscles and ligaments that operate five ever-growing teeth. With these the sea urchin rasps algae off rocks. Even today we call the structure 'Aristotle's lantern'.

An inspection of an animal's insides, however, need not be invasive. Just as doctors in

SPACEWATCH *The long-distance movements of large birds, such as black-browed albatrosses, can be tracked by satellites orbiting the Earth.*

hospitals use ultrasonic scans to examine unborn babies, biologists in research laboratories use the same tool to reveal the way in which organs work. Ultrasonic beams, directed at the complicated system of pipes and flaps in the dolphin's throat and head, for example, are being used to understand the way in which it produces the broad repertoire of clicks, burps, screams, whistles and moans that Aristotle described more than 2000 years ago.

SOPHISTICATED STALKING

For many of these methods to work, the animal under study must be captured, settled into a life in captivity, and brought into the research laboratory – all of which may lead it to behave abnormally. As a result, there is now a growing tendency to study an animal in its natural environment; 'telemetry' has enabled scientists to do just that.

Small radio transmitters and acoustic tags are attached to animals in order to discover where they actually go, and when. A field biologist, sprouting a large array of directional aerials, is then able to follow the tagged animal at a discreet distance and to interfere less with its normal way of life. This technique has confirmed how a male tiger, previously darted with a soporific, collared and released, moves around his territory, revealing that the male's patch overlaps with the territories of several females. It has also allowed researchers to locate their target animals quickly during long-term studies. A pride of lions, for example, may be found only every few weeks after extensive searching, but if a single lion in the pride carries a radio collar, the pride can be located in a matter of hours and therefore observed more often.

The most exciting form of tracking, however, is achieved with the aid of satellites in space. Large creatures, such as whales, basking sharks and polar bears, have been fitted with transmitters that beam signals to satellites in Earth's orbit. The signals are reflected back to receiving stations on the Earth and then on, via land lines, to research laboratories for analysis. Satellite tracking makes for monitoring over far larger distances than terrestrial radio tags can cover, and there is obviously even less likelihood of intruding on the animals under observation.

One of the most remarkable stories to unfold from data collected via space is that of the wandering albatross in the Southern Ocean. Pairs of albatross breed on sub-Antarctic islands, such as South Georgia, but when the single chick has hatched, one of the adult birds must go to sea from time to time in order to find food – mainly squid – for its hungry offspring. During these trips the albatross travels huge distances, taking advantage of the wind to carry it to feeding sites. On its return it avoids flying against the wind, and either tacks back in long zigzags, like a yacht, or loops around until it encounters favourable winds to take it back to its nesting island.

One bird fitted with a transmitter that could be tracked by a satellite in space showed scientists that on each single foraging trip it might travel between 2239 and 9320 miles (3600-15 000 km). The bird, which was confirmed by ground-based observations to be sharing incubation shifts

CHILLED WRIGGLER *An ice-worm moves through an Arctic glacier, a habitat where it thrives.*

OUTSIZED TEDDY BEAR *The Kodiak bear, the largest living carnivore, was discovered and described for science in the late 19th century.*

with its mate, flew at speeds of up to 50 mph (80 km/h) and covered 560 miles (900 km) a day, flying day and night, with few stops, for four days or more. The mate waited patiently on the nest for the wanderer to return.

LIFE AT THE EXTREMES

Through observation, and the technology that has extended and refined it, we have discovered life in the most unlikely places: some spadefoot toads lie buried in hot, North American deserts for years, waiting for the rains to come; the Laysan teal feeds on swarms of flies that breed on the carpet of algae growing at the edge of the remote Laysan lagoon in the Hawaiian islands; and gigantic, 10 ft (3 m) long, red tube worms, and huge mussels and clams, some of them 12 in (30 cm) across, congregate around hot springs at the bottom of the deep sea.

However, not all curiosities are big. There is a greater diversity of life hiding between the grains of sand on an Atlantic beach than is found in a forest. The grains are like giant boulders to the thriving beach life. Entire colonies of tiny creatures, known as bryozoans or 'moss animals', may nestle in a cavity on a grain of sand. There are 5000 known species,

most of which resemble tiny coral colonies or patches of lichens. Each individual or zooid can be specialised for a particular task: there are reproducing zooids, attachment zooids (which attach the entire colony to the rock), defence zooids (which resemble miniature hand grenades) and feeding zooids (with a brush on top that serves as a feeding tentacle). All are so small that you need a powerful microscope to see them. Nevertheless, it is estimated that about 6300 colonies, made up of three or four zooids each, occupy 1 sq yd (0.836 m²) of sand down to a depth of ³/₈ in (1 cm).

Ice and snow may seem impossibly harsh habitats, but even here animals have found a vacant niche and moved in. Marine copepods, distant relations of shrimps and crabs, congregate under the Arctic ice and graze on ice algae. Ice-worms are seen squirming through glaciers, and primitive insects, such as wingless springtails, bristletails and grylloblattids, can stain the snowfields black. In Nepal a new species of flightless, chironomid midge was discovered in 1984 by a group of Japanese glaciologists. The midge remains active at temperatures as low as –16°C (3°F) and lives its entire life in Himalayan glaciers, the coldest home ever recorded for an insect.

NEW DISCOVERIES

Unexpected discoveries are made all the time – not just new observations about the way animals behave and function, but also revelations of new species or even new families of animals previously unknown to science. In 1976 a new species of shark was hauled out of the Pacific Ocean, off the Hawaiian coast, and dubbed 'megamouth' on account of its large mouth. The 14 ft 9 in

NEW SPECIES *The megamouth shark, discovered in 1976, is thought to attract shrimps into its mouth with the aid of bioluminescent spots on its lips.*

(4.5 m) long shark is so unlike any other shark that it was assigned not only its own species and genus – *Megachasma pelagios*, or 'great yawning hole of the open sea' – but also its own family, the Megachasmidae.

Despite all the technology and all the knowledge, however, our way of sampling and studying life on Earth, particularly in the tropical rain forest or the sea, can only be described as 'primitive'. Imagine a visitor from another planet cruising 4 miles (6 km) above the Earth's surface and reaching down for samples of terrestrial life with a large butterfly net on a long piece of string. Is it likely that these explorers, fishing for a representative sample of life, would net anything of significance? Maybe they would catch the occasional insect or bird, and perhaps scoop up a few slow-moving mammals, a handful of bushes, and a few tufts of grass. But with such an ineffective method of sampling, they would not

come anywhere near to discovering the diversity of life on this planet, let alone reaching any understanding of the relationship between plants and animals and the places in which they live. Yet this is how we have been looking for life in the last great wilderness – the oceans.

Even now, who is to say that there are not giant serpents in the sea, monsters in lochs and lakes, leftover dinosaurs in the swamps, and abominable snowmen in the mountains? It is possible that a new species of bipedal ape exists in the forests of western Sumatra. The evidence is mounting for an 'orang pendek' or 'short man'. The creature,

TREE-WALKING *Modern technology enables scientists to study in previously inaccessible places. In French Guiana, a lattice of walkways is dropped on the rain-forest canopy by an airship.*

observed three times in 1994 by a scientist, stands about 4 ft (1.2 m) tall and has red, silky body hair much like that of the orang-utan. It has a bare face, a human-shaped nose, and a gorilla-like head and neck. It feeds on ginger plants, fruits, termites, freshwater crabs, nestling birds, and the rice-and-dried-fish lunches of researchers when it can get some.

Mankind has found, scrutinised, identified, classified and studied an amazing range of animals. In the following pages some of these findings, and some of the extraordinary secrets of the animal kingdom are revealed. Aristotle would have been amazed.

PROGRAMMED FOR SURVIVAL

1

WARM FATHER *The emperor penguin father incubates a single egg handed to him by his partner.*

THERE ARE MORE THAN 1.3 MILLION KNOWN SPECIES OF ANIMAL LIVING ON THE PLANET, AND MANY MORE YET TO BE DISCOVERED. EACH MUST ENTER AN OFTEN HOSTILE WORLD AND SOMEHOW SURVIVE. SOME HAVE THEIR ENTIRE LIFE'S INSTRUCTIONS WRITTEN INTO THEIR GENES, AN INNATE BLUEPRINT FOR LIVING HANDED DOWN TO THEM FROM PARENTS THAT HAVE DISAPPEARED LONG AGO AND WHICH THEY ARE NEVER LIKELY TO MEET. OTHERS AUGMENT INSTINCT WITH LESSONS LEARNED FROM DOTING PARENTS, RELATIVES AND NEIGHBOURS AND FROM THEIR OWN EXPERIENCES. FURNISHED WITH THEIR SCHEDULE FOR SURVIVAL THEY MUST EVENTUALLY GO OUT INTO THE WORLD, WHERE EACH INDIVIDUAL IS VERY MUCH ON ITS OWN.

FOREST OF LEGS *A three-day-old baby African elephant is well protected from predators such as lions and hyenas.*

THE DEBUT

The first few minutes of an animal's life outside the egg or womb are critical. This is the time in its life when it is most vulnerable, to the elements and to predation, but nature has provided the majority of babies with effective ways to survive.

Almost every animal, whether a gigantic blue whale or tiny mite, starts life as a tiny egg. After fertilisation, the egg starts to grow. It is either kept in a safe, warm place in its mother's (or occasionally its father's) body and fed a constant supply of food, or it is wrapped in a shell or a blob of jelly, given a portion of food to keep it going for a while, and either incubated until ready or left entirely to its own devices. By this time the parents have done their best to ensure their offspring's survival, but then comes the time for the youngster to emerge from either the egg or the womb. From the moment they are born, most animals face difficulty and danger, and the first few hours are among the hardest. Parents have adopted all manner of ways in which to give their offspring the start they need to survive.

The first trial for reptile or bird embryos, however, is not the problem of avoiding the array of predators waiting to pick them off as they emerge, but simply that of fighting their way out of the egg. Some reptile eggs have parchment-like shells, and to crack them lizards and snakes have a short, hard, forward-growing egg tooth that is later shed. The eggs of land tortoises and crocodiles, on the other hand, tend to be hard and calcified, like the eggs of birds. In order to pierce the tough shell that has protected it during its early development, the youngster inside possesses a caruncle, a horny outgrowth with which it cracks the egg open before levering itself out. This caruncle, like the egg tooth of lizards and snakes, is eventually shed.

The most common method of exit from a bird's egg – and one adopted by chickens, blackbirds and geese – is known as 'symmetrical hatching': the chick chips a circular cap in one end of the egg and pushes at it; finally, it twists its body in the shell, using its legs and wings to struggle through the hole. Long-beaked birds, such as oystercatchers, avocets and lapwings, adopt a different method, known as 'asymmetrical hatching'. The chicks rotate very little in the egg; instead, they enlarge the hole considerably, slitting the eggshell lengthways, and then escape. A third style of hatching is found among the megapodes, or scrub fowl, of Australasia. Their chicks do not chip at the inside of the shell at all, but stretch and tear at it with their claws, rotating at the same time.

Once free from the confines of the egg, the young bird or reptile faces the problem

THE GREAT ESCAPE *A domestic chick breaks out of its egg by piercing the shell with its 'egg tooth' (top left) and cracking around the blunt end (top right). It then forces the two parts of the shell apart and emerges (bottom left and right), its feathers sticky with egg fluids. Within hours its down will dry and the chick will be running about.*

NESTING IN TERMITE MOUNDS
The female Australian lace monitor lizard (bottom right) is one of several animals that nest in termite mounds. She digs a hole in the wall and deposits her eggs inside. The termites swarm around the hole and repair the damage. The eggs remain walled up inside the mound for the entire incubation period. The termites ignore the eggs, and the lizard embryos develop in the safety of the termite nest.

ST BUILDING *A pair golden-shouldered rrots excavate their st hole in a termite und in April. If the ood is successful, the rents will return the lowing year.*

MOTHER RETURNS *Nine months after egg-laying, when the eggs are ready to hatch, the mother lizard digs a hole in the termite nest. When the baby lizards emerge they will be able to escape.*

TERMITE COMMUNITY
The queen, her consort and attendants are at the centre of a colony containing millions of workers and soldier termites.

of getting out of the nest. When it comes to hatching time, the 12 in (30 cm) long hatchlings of the Australian lace monitor lizard find themselves trapped in a prison. The 7 ft (2 m) long mother uses her powerful front claws to dig into the hard walls of a termites' nest and deposits her eggs inside; and the termite tenants, being diligent on household maintenance, repair the damage. The eggs are safe inside, incubating for about nine months in a warm, humid environment with a constant temperature.

The baby lizards have no chance of scraping their way out of their incubator, and their method of escape was a mystery until an Australian researcher kept watch on several termite mounds that were known to contain lace monitor lizard eggs. His patience paid off, for he discovered that in early spring, at about the time the eggs were ready to hatch and two months before the time for laying new eggs, each nest was visited by a female lizard, most likely the mother of the eggs inside. She dug into the nests, making a small aperture through which the baby lizards could escape.

Crocodile and alligator mothers, too, have to dig their offspring out of a nest, which they build from earth or a mound of vegetation, when their eggs hatch about two months after being laid. The mothers are summoned by the frantic chirping calls of the hatchlings and, having dug them out, they carry them to the lake or river in a special 'gullar pouch' in the bottom of their mouths. A gentle squeeze on a reluctant egg – from powerful jaws capable of ripping prey apart – can help a trapped crocodile hatchling to freedom.

RACE FOR THE SEA

The mothers of sea turtles dig a hole in the sand in which to lay their eggs, covering them over so that the site is completely hidden, and are long gone when the time comes for their offspring to emerge from the eggs. About seven weeks after their mothers have returned to the sea, the baby turtles – as many as 50-150 in each nest – hatch simultaneously. They are on their own, with no parental care, and must make

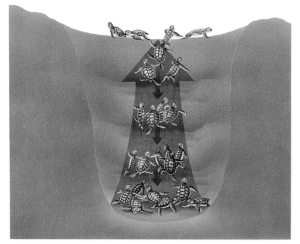

LIFT TO FREEDOM *The 2 in (5 cm) long turtle hatchlings emerge from their eggs and scramble to the surface as if going up in a lift.*

their way to the surface from their underground nests, which can be 20 in (51 cm) down in the sand. They do this by taking the lift. The first to hatch stimulates the adjacent egg to hatch by moving around, and this is repeated throughout the nest. Thus, the hatching of the entire clutch is coordinated. The hatchlings do not dig their way to the surface immediately, but wait until the other eggs have hatched. As more and more emerge, the hatchlings thrash about wildly. The tiny turtles at the top of the nest scrape away at the ceiling, while those at the side undercut the walls, and the ones at the bottom trample and compact the sand filtering down from above.

The cooperation, thought to be instinctive, enables most of them to reach the surface safely. Sometimes the activity stops, and it is up to the shift on the bottom floor to squirm in order to trigger another period of movement. Thus, in a series of fits and starts, the ceiling falls, the floor rises and the chamberful of hatchlings erupts onto the beach. Without this assistance from their nest-mates, the turtles on the lower

DILIGENT MOTHER *The Nile crocodile mother is summoned by the calls of her newly hatched offspring to carry them, in her mouth, to the water's edge.*

RUNNING FOR THEIR LIVES
Green turtle hatchlings make their first journey, from their beach nest to the sea, without their parents' protection.

levels would be unlikely to reach the surface. Once there, however, the hatchlings' problems really begin.

On the beach, sea turtle babies face their first journey. By now the bonds between nest-mates have loosened, but a band of hatchlings moves more rapidly across the open beach than a solitary individual, and straying or lethargic babies are pushed back on the straight and narrow by the rush of the others. After a few false starts, most youngsters point accurately towards the sea and can orientate themselves correctly whatever the weather, during the night or by day, and with or without cloud cover. It is not certain how they find their way, but the quality of light over the sea may be a guiding factor.

The young turtles must race for their lives, their tiny flippers acting like oars to propel them over the sand. The trip may be short, perhaps only 100 yd (90 m), but it is one of the most dangerous that they will make during their lives. Sea turtles rarely emerge from their nests during daylight hours because the sun and sand are too hot. And

at night they have a better chance of avoiding the monsters on the beach. Those that do emerge when it is still light are confronted by one threat after another. On the Great Barrier Reef of Australia, green sea turtle hatchlings are picked off by feral cats, dogs and pigs, dingoes, foxes, large monitor lizards known as goannas and by the opportunist silver gull, which drops in to take advantage of the feast.

In February and March each year, thousands of nankeen night herons, whose breeding period coincides with the hatching of the turtles, congregate on Raine Island, at the north end of the Great Barrier Reef. Every evening the herons patrol the beach, grabbing hatchlings as they emerge. As the exodus starts, the birds make the walls of the egg-chambers collapse by stamping their feet on the sand and pecking at the nests, and the young turtles are trapped and picked off one by one.

Further down the beach, towards the sea, the ghost crabs come out from their holes in the damp sand and grab the baby turtles in their huge claws. If they catch the rounded turtle shell, the baby may be able to wriggle free, but a firm pincer-hold on a flipper seals the turtle's fate. The baby turtle is dragged down into the crab's burrow and devoured at leisure. Meanwhile, other crabs move rapidly across the sand, scanning the beach with their periscopic eyes, but they are wary of trespassing on another's territory and do not run far from their burrow entrances.

Many hatchling turtles slip through the line of crabs, but on reaching the tideline they are at risk from a different predator, one that swoops aerobatically down from the sky. Frigate birds can pluck baby turtles from the beach without touching the ground, and

remain on the wing as they swallow them. Meanwhile, in a tide pool a waiting octopus, well camouflaged with a covering of algae, puts out a long, suckered arm and draws its victim beneath an extended umbrella towards its shell-cracking beak. Even

---

BIRTH RANGE

The blue whale, the largest animal ever to have lived on Earth, has a gestation period of 11 months and gives birth to the largest baby – weighing 3 tons (2700 kg) and measuring 24 ft (7 m) in length. The African elephant, the largest land animal, has a gestation period of 22 months, and a new-born baby weighs 265 lb (120 kg). The shortest gestation period known is just eight days, for an American opossum, although 12 days is more usual. The largest litter – 31 babies – has been attributed to the tailless tenrec of Madagascar. And the smallest mammal babies are those of the mouse opossums of Latin America, which are no larger than a grain of rice.

---

for those hatchlings that reach the sea, the dangers are not over. Black triangular fins cutting through the water just beyond the surf line signal the presence of sharks. Next
continued on page 20

SEASONAL FEAST *A gull waits for a turtle hatchling to emerge from the sea. Overleaf: With few predators about, albatross chicks sit on raised nests of mud and vegetation on the ground.*

morning, banded land rails and silver gulls scavenge among the debris.

The survivors swim about 8 in (20 cm) below the surface, coming up for a gulp of air every five to ten seconds. They are vulnerable not only to fish, but to any aerial predator soaring overhead that can crash-dive to a depth of 10 ft (3 m). Despite the predation, however, millions of baby turtles survive the first few hours. They are nourished by the remains of their yolk sac, a leftover from the egg, until they reach the ocean currents, where they feed on the drifting life around them. They swim to deep water, travelling directly away from land and correcting their course each time they surface for air.

The carnage at dawn and dusk is counterbalanced by the number of turtle babies born at a time. Each turtle mother produces hundreds of eggs and deposits them on the beach on several separate occasions throughout the breeding season. With so many hatchlings emerging at the same time predators are sated, and some turtles are bound to survive.

With the exception of the flat-back turtle hatchlings, which stay near the shore, none remains near its place of birth, and no one has established for certain where they go during what is known as 'the lost year of the turtle'. It is thought that they join the surface plankton and move with the ocean currents, feeding on the tiny plants and animals that drift with them, until they grow to the size of dinner plates and return to the shallow-water breeding grounds of the Great Barrier Reef.

FAMILIAR WORLDS

Female sea turtles journey immense distances – up to 1400 miles (2300 km) – under water to the place where they were born, in order to give birth themselves.

How they find their way and recognise their own beach is still something of a mystery, although it is thought that the Earth's magnetic field offers a navigational clue and the smell of the sand on the beach identifies the nest site.

Baby turtles in a seawater-filled laboratory tank have been fitted with miniature swimsuits that are attached to monitoring devices and then subjected to different magnetic fields. By altering the direction of the magnetic field, scientists have seen that turtles respond to some component in it, most probably the angle of inclination – the angle at which the magnetic field radiates from the Earth, which varies at different points on the globe. Turtles in the sea have also been fitted with radio transmitters and followed closely in order to reveal this baffling, long-distance homing behaviour.

The female lemon shark also returns to her point of departure to breed. She does not deposit eggs but, like mammals, gives birth to live young. One of the lemon shark's birth sites is the Bimini Atoll, not far off the Florida coast in the Bahamas. Here, a tangle of mangrove roots fringing the shallow horseshoe-shaped lagoon provides a nursery for the young sharks. Directly after

A SAFE LAGOON Lemon sharks at Bimini Atoll are born, not hatched. A few seconds after emerging from the womb, the pup swims to the safety of the mangroves, where it will spend the first few months of its life in its own narrow strip of territory.

BABY GIANT *Born tail first, the southern right whale may be pushed to the surface by its mother to take its first breath. It dives below her between breaths to nurse and remains close to her for about 14 months.*

birth, often with the birth membranes still attached and dragging behind them, the newborn sharks swim immediately for the safety of the mangroves.

During their first year, the baby sharks patrol a 1300 ft (400 m) long by 130 ft (40 m) wide section of mangrove shoreline, feeding for much of the time on small fish, such as snappers and grunts (so-called because they make grunting sounds), and occasionally on invertebrates, such as shrimps and worms. During one period in their growth, when they are between 30 in and 40 in (76-100 cm) long, they develop a passion for octopus – if they can catch them.

As they grow older, the sharks gradually enlarge their field of activity until there comes a time, at about the age of two, when they leave the nursery area and move to other sites within the lagoon. At each site, only sharks of the same age and size associate together. They remain inside the lagoon until they are about seven or eight years old, and then head out towards the more open reef habitats. At about 11 or 12, the lemon sharks move away from Bimini entirely, making long migrations up and down the US coast and working the reefs down to a depth of 165 ft (50 m). The males move as far north as Virginia, returning to the Keys in Florida Bay in May, June and July to mate. The females have a two-year reproductive cycle. They mate, carry to term and drop their pups in one year, and then have a year off. Pregnant females only return to Bimini to pup.

Pregnant southern right whales also head for home – special sheltered or protected bays in the Southern Hemisphere – when the time comes to give birth. As the moment of birth approaches, the mother stays close to the surface, sometimes with her tail and head arched out of the water; her breathing rate increases and she defecates frequently. As the amniotic membrane ruptures, many pints (litres) of birth fluids and blood escape into the water and the baby whale is delivered tail first – the entire process taking up to an hour for an easy birth and maybe four hours for a difficult one. During the birth the baby's fins lie flat against depressions in the side of its body, where they cannot become tangled in the umbilical cord. The cord itself is short and thick. It is not severed by a bite from the mother, but breaks at special points of weakness when it is snapped taut.

With no air in its lungs the baby tends to sink, but instinctively it swims for the bright surface, sometimes helped by its mother, who nudges it upwards with her snout or

BURSTING WITH LIFE

The tiny female sea louse, *Paragnathia formica*, just ¹/₈ in (3 mm) long, pays the ultimate price for motherhood. She begins life as a larva feeding on the blood of fishes, but after the third feast and a third moult she joins her tiny, ant-sized mate in his mud burrow and devotes the rest of her short life to her offspring.

The male is well-armoured and a diligent father, protecting his partner and the offspring growing inside her. The mother is blind, has no sense organs with which to find her way about, no appendages with which to move, and no mouthparts for feeding. With pregnancy, she becomes a baby factory. The youngsters developing inside her take over her entire body, causing her internal organs to degenerate. Offspring are packed into every conceivable corner and are clearly visible through her translucent cuticle.

When the offspring are ready to emerge into the

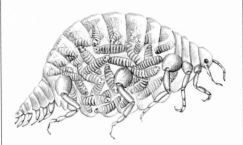

world, the male sea louse breaks down the mud at the burrow entrance, and the rising tide pours in. Immediately, the mother sea louse absorbs water and expands, until eventually she bursts along a line of weakness in her cuticle, just beneath her head. The babies rapidly swim free, leaving her empty skin behind. She has paid for the debut of her offspring with her own death.

fins. The newly born whale takes its first breath about ten seconds after leaving the womb and will be swimming competently within a half an hour. Ten hours later, the afterbirth is expelled.

Giraffes, too, have traditional birthing sites. After a 15 month pregnancy, female giraffes abandon their nomadic life on the African savannah and travel to the same places that mothers-to-be return to year after year. There might be several such sites

within a female's home range, but the giraffe's fidelity to a single site is so strong that one giraffe in the Serengeti, with a home range of 37 sq miles (95 sq km), dropped her second calf within 50 ft (15 m) of the place where she had given birth to her first.

At these calving sites, several baby giraffes are born at roughly the same time, although each newborn calf, with a height of about 6 ft (1.8 m) and weighing 15 stone (95 kg), is born in isolation. A calf must quickly get to its feet, bring its long, swaying neck and gangly legs under control and learn to run before predators, such as lions and hyenas, have picked up its scent. It is curious that predators have not yet caught up with the fact that giraffes are so predictable. If they had done so, the giraffe's calving behaviour would perhaps be more random.

An advantage of synchronous calving at traditional sites

FREE FALL *The baby giraffe drops 7 ft (2 m) to the ground at birth. After a good licking from mother, during which the bond between them is forged, the baby is up and running before the predators arrive.*

ANIMAL MIDWIFERY

Some animals do not give birth alone, but have midwives in attendance. Elephant 'aunties', young female relatives of the mother, are quick to help a newborn calf to its feet, while in a similar way helpers may assist a baby dolphin or whale to the surface to take its first breath. The most extraordinary birth ritual, however, must be that of the Rodrigues fruit bat.

The bat is a seriously endangered species; its isolated population on Rodrigues Island, near Mauritius, was down to just 350 individuals in 1991. Every few years cyclones hit the island, killing many of the bats, and survivors can do with every bit of help they can get – from each other.

When a new mother is about to give birth, for instance, some of her female neighbours might gather around to help. A midwife licks the mother's vagina and then the baby's head when it first appears. The midwife also gives advice. She demonstrates the feet-down position in front of the first-time mother, hanging from the claws on her thumbs and encouraging the mother to adopt this position, rather than the normal head-down position, for the birth. The midwife also fans the mother and embraces her with her leathery wings.

A difficult birth may last for several hours. Throughout the process, male fruit bats place themselves between the delivery party and the outside world, protecting mother and offspring from possible predators, such as tree snakes.

EVERY BABY COUNTS
The very rare Rodrigues fruit bat mother is helped by her neighbours when the time comes to give birth.

is that young calves can be left by their parents in a group that is looked after by one or two adults. In the heat of the day, when predators are resting, 'nursery' groups of six to twelve calves, none more than a couple of months old, are a feature of giraffe social life. Giraffe mothers and babies have a tendency to stray from each other in a way that rhinos, for instance, do not. Their long necks and good visual contact make it relatively easy for giraffe parents to spot the nursery groups and return to feed their offspring. When danger threatens, giraffe babies first hide between their mother's legs, underneath her belly. The calf stands at such an angle that the mother is able to kick out at a predator without injuring her baby.

At two to three months old a calf picks at bushes and leaves; it ruminates at three to four months old, and is not fully weaned for a further twelve months. All the while it must follow the constant migration of the herd, and newborn giraffes have been seen walking beside their mothers when they are only an hour old.

The echidna or spiny anteater baby is even more dependent on its mother. Although a mammal, the female lays an egg into a pouch, and when the tiny baby hatches ten days later she feeds it on milk produced by glands in the pouch. The baby remains in the pouch for three months, where it must hold on tight to avoid falling out, until its spines erupt and afford it some protection from marauding predators, such as lace monitor lizards.

THE ECHIDNA ENIGMA This young Australian echidna, or spiny anteater (right in picture), is a mammal, yet it started life by hatching out of a soft-shelled egg like a reptile.

DEPENDABLE PARENTS

An animal's formative years might be spent with mother or father, or with both parents. Parents provide a safe place in which their offspring can grow up, enabling baby to acquire all the knowledge and skills it needs to live without them.

Parents and offspring in the natural world are often difficult to spot, and good fieldcraft is essential if a scientist is to uncover the secrets of an animal's early life. It pays an animal not to advertise or draw attention to itself for that could attract the attention of unwelcome guests, so parents are careful where they situate their nests or home sites. In order to survive, a young animal must blend in with its background, hide in a safe and secret place, or be protected by a fierce or formidable parent. And while some young animals are equipped to fend for themselves almost from the minute they are born, many depend on their parents for food and protection: for as long as two or three years in the case of geese and swans, and even longer for many large mammals.

Most birds hide their young in nests, ranging from a hollow in a hippopotamus dropping – the nest of the water dikkop – to an elaborate construction at the top of a tree like that of the African hammerkop, to a long tunnel laboriously scraped out by beak and feet like that of the European kingfisher. Many different methods are used to protect eggs and young from predators. The fairy tern, from the tropical Pacific, deposits its egg precariously on the branch of a tree, thereby avoiding nest smells that might attract predators. The marbled murrelet of North America builds a moss-lined, cup-shaped nest of guano at the very tops of the highest trees in the world. Nobody had ever seen the nest of this species until 1974, when the first marbled murrelet nest was discovered 148 ft (45 m) from the ground at the top of a Douglas fir in California.

CREATING A COMFORTABLE NEST

The simplest nest, or non-nest to be more precise, is probably that of emperor and king penguins. They use no material at all, not even a scrape in the sand. During the icy cold Antarctic winter, when temperatures drop on average to –55°C (–67°F) and sometimes to –89°C (–128°F), making it the coldest place on Earth, the male emperor incubates a single egg by balancing it on his webbed feet, clear of the chilling ice and snow. Holding it close to his belly brood patch under a thick fold of skin, the emperor incubates the egg at about 37°C (98°F) – up to 80°C (176°F) warmer than the temperature of the air outside. When the chick hatches, it sits on its father's feet in order to keep off the cold surface ice.

For three species of geese in the Arctic, a major problem is that of keeping their chicks warm. The amount of down used by emperor, Canada and brant geese for lining their nests depends on whether the nest is close to other nests or whether it stands alone. The geese balance the advantages of lining their nests and keeping the eggs warm against the disadvantage that pale-coloured down and light-coloured eggs

WARM ARCTIC NEST *Canada geese parents are able to desert their nest occasionally and stretch their legs as the eggs are protected from the cold by a thick lining of down, mixed in with vegetation for camouflage.*

might attract predators such as skuas and glaucous gulls.

Brant geese pack their nests with down, and incubating mothers leave the nest more frequently than the other two species in order to stretch their legs and preen. Since brants nest in colonies there are always tolerant and attentive neighbours to chase predators away. Emperor geese, on the other hand, nest alone, so mothers sit tight, rarely venturing far from the nest for fear that their eggs might attract unwelcome attention. With mother geese incubating their eggs for most of the time (they leave the nest for brief periods of not more than 13 minutes, on alternate days), there is little need for a downy lining. Unlike brant geese, they cannot leave: their nests are too conspicuous and, unguarded, they would be wide open to predators. Canada geese, which tend to nest in loose colonies but are highly territorial, have found a compromise:

they leave the nest occasionally, so they line it with down to keep the eggs warm, and camouflage it with vegetation.

Different manoeuvres are required in order to keep chicks cool: herons defecate on their nests to promote cooling by evaporation, while the white-crowned black wheatear of the Sahara constructs its nest on a 'cooling tower' of pebbles. The small stones trap dew, which evaporates to cool the nest cup perched at the top. Sand grouse mothers wet their breast feathers in order to cool their eggs, while other birds living in hot climates shield their eggs and chicks from the heat of the sun with their body and wings.

Chicks in a nest are not always treated equally. Whereas the mothers of eastern

SOLID TREE HOUSE *The roofed nest of the African hammerkop can be up to 6ft 6in (1.9m) wide and contain 10 000 sticks plastered together with mud and straw. Below: The fairy tern of the Pacific islands lays its single egg on a bare branch.*

THE PHYSIOLOGY OF PLAY

A lot of research has been done on animal play and there are many theories. A baby lemur spinning on the spot, a red deer fawn frolicking in the river or a brown bear cub plucking a flower and racing off across the meadow may seem inexplicable until it is understood that these are all examples of young animals indulging in a very 'human' activity: play.

As with human children, play is thought to contribute to an animal's physical and mental growth. Why else would pronghorn fawns and Norwegian rat pups spend 20 per cent or more of their valuable calories on play, such as leaping about for no apparent reason at all? They would not use up vital energy, or expose themselves to predators and to dangerous sites such as treetops and water, if play were not an essential part of growing up.

Play occurs at a critical stage in the development of an animal's muscle and nerve fibres, and periods of the greatest playfulness coincide with the active formation of connections in the cerebellum, that part of the brain which regulates movement and posture: as the animal learns something new, the connections are modified. Other parts of the brain also

benefit from play, particularly in young primates and dolphins, which have large, well-developed brains and continue to play at an older age than is the case with other animals.

Play also seems to influence the formation and coordination of muscle fibres, by sending nerve signals to the young animal's developing muscles. Here, too, researchers have noted that the greatest period of fibre formation coincides with the peak time of play. Play also enables young animals to rehearse skills that they will need to survive as adults. Predatory animals learn hunting skills – stalking, pouncing and a quick kill – while animals attempting to stay off their menu indulge in mock flight, bolting from invisible predators, squirming and kicking, bouncing and running.

The most unlikely animals indulge in play. The young great anteater of South America practises bluff charges; it raises its hackles, lifts a foot, hops sideways on its other three feet, all the while roaring furiously. Its curious behaviour is thought to deter predators and to discourage other anteaters from approaching its favourite termite's nest. The hatchling sea turtle also plays, holding up a front flipper

and vibrating it in the face of a nest-mate – a gesture it might use many years later when it returns to the offshore waters of its birthplace during courtship.

Social animals find their status in society through play, and potentially aggressive members of society are brought under control during games. Bouts of play help to ease an animal's passage into the life of the group, ensuring that it becomes a well-integrated member of society. Close studies of two species of monkey, squirrel monkeys and rhesus macaques, have shown that, from about three months old, the youngsters spend half their waking hours playing.

Peccaries play on and off for most of their lives. The aggressive collared peccary, for example, plays several times a week; herd members snap at each other, rolling and locking jaws. Play in peccaries encourages group cohesion, and bouts stop just as suddenly as they begin – probably because of some hidden olfactory (smell) signal emitted by one or more of the herd. Play signals such as this are vital. In order to avoid real conflict, one animal must tell the other that it

THE HUNTER *A baby North American mountain lion, or cougar, practises its hunting techniques on a desert tortoise.*

THE HUNTED *A baby springbok hones its developing muscles and reflexes ready for the moment when it must flee from predators.*

is not attacking but playing. Domestic dog puppies do this by crouching forward and raising their rear ends in the air; rats flip onto their backs, and so does a species of parrot, the New Zealand kea, whose play is more like a primate's than a bird's.

Animals can also change their play behaviour depending on the circumstances. In an experiment at the New York State Psychiatric Institute, a laboratory rat was invited to baby-sit. The young three-week-old rat, the equivalent in developmental terms of a seven-year-old child, was placed with a litter of helpless, pink newborn pups. At first, the youngster tried to play, attempting to pounce and wrestle with them as it would with a nest-mate. After a few days, however, its behaviour changed. It would gently gather the pups into the nest, retrieving any which crawled away.

Animals like the young rat appear to be able to adapt various aspects of their behaviour to their circumstances – in this case, shifting from the rough-and-tumble behaviour of a child to the serious maternal or paternal behaviour of an adult.

bluebirds in North America treat all their children the same, regardless of sex, fathers give their daughters more attention and twice as much food. This is probably because the male offspring will soon be looking for territories near by and may become a threat to their fathers, whereas daughters move to a new area to breed. Some predatory birds, such as eagles, owls, bee-eaters and herons, stagger the incubation of their eggs so that the chicks hatch at different times. If there should be a sudden food shortage, the older chicks are fed in preference to the younger ones; if times are really hard, the younger chicks may be left to die and even utilised to feed their older siblings.

SITE SELECTION

Birds are not the only animals to find safe nest sites in trees. Some very unlikely parents go to great lengths to protect their young. It might seem strange to find a crab nesting in a tree, but the bromeliad crab of the Jamaican rain forests does just that, providing a nursery and food for the young. The Jamaican highlands are saturated with a daily downpour of rain, but very little of this accumulates as standing water. This poses a problem for the crabs as they need a pool in which to reproduce, and so the mother crab climbs to the forest canopy carrying her developing eggs under her tail.

She finds the tiny pools of rainwater filling the centres of bromeliad plants that grow on the branches of high forest trees, and clears the water of leaf litter and other debris that would decay and deoxygenate the water. She does, however, leave snail shells in the pool because these provide additional calcium for her young and reduce the water's acidity. When the eggs hatch, the mother supplies her offspring with snails, millipedes, cockroaches and other insect morsels.

SCORPIONS, SPIDERS AND INSECTS

Not all animals depend on a nest site. Those with a formidable armoury, such as the mother scorpion, dispense with the need for a nest altogether. The scorpion carries her youngsters – miniature replicas of herself – in a great writhing mass on her back. They are protected from any bird predators by the female's poisonous sting.

BABY BACKPACK A tailless whip scorpion mother keeps her young safe from the predators of the Amazon rain forest by carrying them on her back.

BABY SLING A wolf spider, the wandering spider mother from Costa Rica, carries her egg sac slung under her body.

Baby spiders are no less vulnerable than the young of many other animals. Wolf spiders carry their eggs under the abdomen, in an egg sac that can be almost as big as the mother herself. After two to three weeks the mother bites open the egg sac, and her brood of 100 or more climb aboard her body, sometimes several layers deep. Here, they feed on the reserves of yolk, holding on for another week or so. Nursery web spiders also carry an egg sac, but in the mouth. When it is time for the young to hatch, the mother places the sac on a stem or leaf and surrounds it with a silken tent. She then stands guard until her youngsters have undergone their first few moults and are ready to face the world alone, at which point they disperse.

Some spiders, like birds, feed their offspring regurgitated food. Theriodion spiders spin three-dimensional webs in hedges, with the female creating a thimble-like shelter for herself in one corner. Here, she rears her brood, at first feeding them a

continued on page 30

THE ANT AND THE BUTTERFLY

The European large blue butterfly has been exercising the minds of British entomologists because, although it was declared extinct in Britain in 1979, until recently nobody knew why it had disappeared. The reason became evident, however, after specimens brought from the European mainland were kept and observed in captivity. Their life cycle, it turned out, was inextricably linked to a red ant, which in turn was influenced by the common European rabbit, sheep, and a common downland plant, thyme.

Little was known about the butterfly's life history until 1915, when Bagwell Purefroy and F.W. Frohawk tried, but failed, to breed it in captivity. The adult female laid her eggs on thyme buds and, a week later, tiny caterpillars hatched out and fed, sometimes on each other but more often on the thyme plant itself. When the caterpillars changed their coat for the third time, becoming what is known as third instars, they simply stopped feeding and died. Purefroy and Frohawk were perplexed, although

AROMA OF CATERPILLAR *The female ichneumon wasp (right) uses her keen sense of smell to locate the caterpillars of the rebeli blue butterfly in a red ants' nest. Below: The life cycle of the European large blue butterfly is dependent on the cooperation of a colony of red ants. The ants guard the caterpillar and pupa in exchange for sugar, and escort the emerging adult from the nest.*

Purefroy believed he had found a clue when he noticed a red ant carrying a caterpillar back to its nest. Many years later, in 1976, the rest of the story emerged.

In the autumn, when the large blue caterpillar reaches its third instar, it changes from a herbivorous diet to a carnivorous one. It climbs down from the thyme plant, and wanders around on the ground until it encounters a red ant of the species *Myrmica sabuleti*. The caterpillar rears up and produces a drop of 'honey' from a special nectary gland on its abdomen, which excites the ant so much that, instead of biting or stinging the caterpillar, it grasps the caterpillar in its jaws and carries it into the heart of the ants' nest. The ants in the colony do not attack, as they would other intruders, but are appeased by the production of this sugary solution at the caterpillar's rear end; the caterpillar also releases pheromones, or

chemical messengers, to suppress the natural tendency of the ants to tear it to pieces. The ants crowd around and 'milk' the caterpillar, palpating it with their antennae to stimulate more secretions. It repays its

hosts by eating their larvae: the ants, however, still do nothing to harm it. In this way, the caterpillar is afforded personal protection in the nest while continuing to fatten up for the next stage of its life cycle.

The large blue caterpillar remains in the ants' nest, feeding on ant larvae and pupae, for about five to six weeks. It then hibernates in the ants' nest until the following spring, when it changes into a pupa or chrysalis, continuing all the while to placate its hosts with sugary solutions and pheromones exuded through the pupal case. By scraping the tip of its abdomen against the inside of the pupal case, the metamorphosing butterfly inside makes peculiar rasping sounds, similar to the sound communication signals being exchanged between the ants themselves, which they generate by rubbing their back legs together.

In spring, towards the end of the pupal stage, the large blue becomes particularly noisy, attracting many ants to gather round. As the adult butterfly emerges, the ants do not attack it, for it too produces secretions that both excite and pacify the hosts. The large blue then heads for the surface through the ant tunnels, with its wings folded, taking with it an escort of ants that it continues to supply with sugar solution in return for the protection that the ants provide against small predators, such as ground beetles.

Another species of blue butterfly, the very rare European rebeli blue, strikes up a more subtle relationship with its own species of red ant. Instead of eating its hosts' progeny, the caterpillar encourages the ants to feed it in the same way as they would feed their own larvae.

The rebeli blue, however, is not completely safe when it is taken into the nest of its ant hosts; it is still at risk from a parasitic wasp that lays its eggs inside the caterpillar and whose larvae then eat the caterpillar from the inside out. The curious thing is that the wasp does not attack the caterpillar on its food plant – the cross-leaved gentian – but pursues it inside the ants' nest. Here, the ants attack the wasp, 20 or 30 at a time biting her wings, legs and antennae. Why should the wasp go to such trouble?

The wasp only shows an interest in nests with the right sort of ants, and only enters nests where caterpillars are present – attracted, it is thought, by their smell. Since 95 per cent of caterpillars end up in the wrong nests and are therefore adopted by the wrong species of ant and die, the wasp ensures the survival of her own offspring by waiting until the caterpillars are already in the correct nests. The female wasp herself is heavily armoured and can resist the attacks for long enough to lay her eggs. But what of the newly emerging wasps? How do they avoid the inevitable onslaught?

Ten months after the eggs have been laid and the wasp larvae have consumed the caterpillar, they pupate, changing into

GRUB MILK *Tropical ants encourage a riodinid butterfly larva to secrete a sugary solution from glands on its body.*

adult wasps. When they emerge from their pupal cases, however, they release a chemical that causes the ants to fight among themselves. So, while most of the colony are brawling, the young wasps make good their escape.

In Central America, the riodinid butterfly, *Thisbe irenea*, is also endangered by parasitic wasps and, in order to deal with them, has developed a close relationship with a species of ant that lives on Croton trees. The worker ants normally collect sugary secretions from the extrafloral nectary at the base of each leaf, but the caterpillar tempts them away with a more attractive solution than that of the tree. About 33 per cent of the tree's secretion is sugar, whereas the caterpillar's consists mostly of amino acids with little sugar; as a result, the ants obtain a far more nutritious meal from the caterpillar.

Normally the ants would visit the leaf nectaries and then return to the nest, but when they visit the caterpillars they tend to stay for several days. A pair of brush-tipped glands on the caterpillar's head produces a pheromone that keeps the ants' attention focused on the caterpillar. And in case that is not enough, the caterpillar has a third organ – emanating from the first thoracic segment and extending over the head – that consists of movable, rod-like appendages. These are brushed against granulations on the top of the head to create 'ant-talk' vibrations, which travel through the leaf or stem and are picked up and understood by the ants.

The caterpillars of riodinid butterflies need their ant guard to protect them from the wasps, which otherwise kill the caterpillars, cut them up and transport them back to their nest. If the wasps should attack, the caterpillar releases the ant alarm pheromone, and the ants immediately assume a defensive posture, driving the wasps away.

With such an intricate life cycle, the large blue butterfly, the rebeli blue and their relatives are vulnerable to any major changes in their environments. The large blue, for instance, is usually found on chalk downland, where the grass is kept short by sheep and rabbits and where thyme plants thrive. These conditions are also essential for red ants. When myxomatosis hit the rabbit population of the British Isles in 1953, the grass grew long. Changes in agricultural practice meant that sheep were removed from the downlands, the vegetation changed, the ants disappeared, and so did the large blue butterfly.

mixture of predigested food and her own intestinal cells, the spiderlings begging her to regurgitate food by stroking her legs. As they grow older, she brings back prey intact, punctures it with her fangs and encourages them to feed for themselves.

The female tsetse fly ensures her offspring's survival in a very distinctive way. Instead of laying many eggs, as other flies do, she retains a single egg inside her body. There it develops, and when it hatches she 'lays' the larva, or maggot, on the ground, where it secretes a pupal case or chrysalis, inside which it changes – a process known as metamorphosis – into the winged adult.

FROG MOTHERS AND FATHERS

Other animals that metamorphose from an immature, non-breeding phase into a fully mature adult capable of breeding are the amphibians, and most, but not all, show

LIVING RUCKSACK *The eggs of the Surinam or pipa toad are embedded in pockets in the skin of the female's back.*

very little parental care. Once European common frogs and toads have laid their eggs in clumps or strings surrounded by a protective jelly, they have little more to do with their developing offspring. The jelly provides some protection from aquatic insect predators and from mould growth, but otherwise the offspring are on their own. Several species of frogs and toads, however, are more attentive to their offspring.

Salamanders guard their eggs, the males usually standing sentry duty, although, in the case of the arboreal salamander, the female coils her tail around the eggs and prevents them from being infected with fungi. The female pipa toad has a curious way of looking after her brood: the male pushes her eggs into the tissues of her back, where they develop in little pockets into tadpoles, and remain until they turn into froglets.

Midwife toads, too, have an elaborate way of looking after their young. The female

SOLITARY CONFINEMENT
The spotted hyena mother spends the first few days after giving birth alone with her newborn cub so that it learns to recognise her.

lays her eggs in a triangular-shaped space formed by her back legs. The male fertilises them, and then puts his legs into the egg mass so that it sticks to his lower body. He carries the eggs for several weeks, ensuring that he is always in conditions with the right temperature and humidity for the eggs to develop. At hatching time he heads for the pond, places his back legs in the water, and the tadpoles hatch out and swim away.

SAFETY FROM PREDATORS

Many other animals also go to great lengths to ensure that their offspring have a good start in life, but, like frogs and toads, without necessarily having a permanent nest site or den. A cheetah mother may have to move her cubs regularly to avoid the attention of lions or hyenas. The hyena mother spends the first week after the birth of her babies in solitude, away from her clan. Although a communal den would be safer, she prefers her own den at first in order to make sure that she has exclusive contact with her offspring and that they become familiar with her. Once the process is complete, the family joins the rest of the clan.

Gazelles, often the prey of hyenas, leave their fawns in the grass, returning every now and then to feed them. The newly born fawn probably has a greater chance of survival in hiding at first than if it followed the herd. Siberian crane chicks, easy prey

THE MILKY WAYS OF THE ANIMAL WORLD

Milk may strike you as a peculiarly mammalian sort of nourishment. Indeed, humans are one of the few animals that continue to drink milk, albeit the milk of another species, throughout their lives. However, the production of a liquid food primarily to feed a newborn or immature youngster while it is still in the tender care of its parent, is not exclusive to mammals.

There is, for example, pigeon's 'milk' – a secretion that emanates from the crop of adult pigeons and doves. Both sexes are capable of producing the thick, protein-rich substance, which is more like cottage cheese in consistency than milk. Low in calcium and carbohydrates but high in sodium, essential fatty acids and vitamins A, B and B_2, it ensures that young birds receive sufficient nutrients to sustain their fast growth rate.

This milk is produced in much the same way as milk in mammals. About half way through the period of egg incubation, prolactin from the pituitary gland stimulates the crop walls to thicken, and they become gorged with blood vessels. Cells lining the crop become milk cells and are sloughed

FISH MILK *Cichlid fry feed on a nutritious, milk-like slime that the mother secretes over the surface of her body.*

off and regurgitated for the chick. To begin with, these cells are generated when the crop is empty and the chick receives pure crop milk; later in the chick's development, however, the milk is mixed with more familiar pigeon food.

Pigeons are not alone in producing crop milk: flamingos and emperor penguins do it, too. Greater flamingo parents produce a crop milk that has less protein and more fat than pigeon's milk, to which they add 1 per cent red blood cells and a liberal amount of canthaxanthin, the pigment that colours a flamingo's feathers pink, and which the bird obtains from the shrimps on which it feeds.

The production of flamingo milk is triggered by the hormone prolactin, but in this case it is stimulated by the begging activity of the chick. Both sexes have the ability to produce the bright red milk, and adults without young sometimes foster and feed orphaned chicks. The habit is thought to be the result of nesting and feeding sites being so far apart.

Emperor penguins have a similar problem. Throughout the bleak months of the Antarctic winter, the male bird balances and incubates a single egg on its feet. The female, meanwhile, is away at sea feeding on squid and fish. Hatching is timed to coincide with the return of the female, filled with partially digested food for the chick. Sometimes, however, the female is delayed and the male has to provide for the chick. He produces a crop milk that is so rich in proteins, fats and carbohydrates that the chick can double its weight in two days – a remarkable feat considering that the father has not fed for over two months.

It is one thing for warm-blooded birds to produce milk, but how about bony fish and sharks? Midas cichlids from the freshwater lakes of Nicaragua – in common with other cichlids – are just as assiduous in their parenting as are mammals and birds. A pair of these fish might deposit and fertilise

up to 2000 eggs. The emerging hatchlings are gathered into a nest scrape on the bottom of the lake and the parents stand guard. At first each little mouthless fish feeds on a grain-sized yolk sac contained within its tiny body, but when the nutrients are used up and the mouth is formed they turn to an unusual food source: five days after becoming free-swimming fry, they graze on a slime secreted by the parent fish's body.

In the tropical and temperate oceans of the world an even more remarkable story has come to light, concerning the tiger shark. Unlike the cichlid, the tiger shark mother retains her embryos inside her body, with as many as 40 pups developing at any one time, each feeding mainly from its own yolk sac. But in addition to this, the offspring of the tiger shark also receive a food supplement from their mother in the form of a creamy milk,

CROP MILK *A healthy greater flamingo chick receives about $1/3$ pint (200 ml) of milk from each parent every day for two months.*

which is secreted from the wall of the uterus. In this way, the mother tiger shark ensures that her many offspring grow fast and well before they emerge from the womb to fend for themselves in the vast and dangerous sea.

Among mammals, the production of milk, in whatever form or guise, is usually associated with the female of the species, but a recent discovery has revealed a male mammal that produces real milk: the male Dayak fruit bat from Malaysia. However, while sexually mature males have been found with functional testes and mammary glands, it is not known whether fruit bat fathers suckle their young or whether these glands are an evolutionary oddity.

for Arctic foxes and skuas, also hide on the ground at first when they hear their parents' alarm calls, but later resort to running and swimming for their lives. The fry of the eel-like arawana fish from the Amazon hide in their father's mouth when danger threatens, as do the mouth-breeding cichlids of African lakes.

ON THE MOVE

One way to keep an eye on a baby and ensure that it keeps up with the group is by carrying it. Giant anteaters and koalas carry their youngsters on their backs, while young monkeys hang from their mother's tummies. Chimpanzee babies start life on their mother's belly, but rapidly progress to her back and shoulders, where they get a good view of the countryside passing by.

Some youngsters have a hard time keeping up with mother: the baby wildebeest, for example, has to be on its feet and following the migrating herd within moments of its birth – or it will fall prey to hyenas or lions. Some of the most extraordinary tales of animals keeping up with their parents come from those precocial birds (birds that are active after hatching) that must find food for themselves as soon as they hatch. There is very little food for wood ducklings in their nests in the hollows of trees, so they have to leap into space, falling 20 ft (6 m) or more to bounce on the leaf litter below. Their first 'flight', however, pales into insignificance beside the exploits of the barnacle gosling. Barnacle geese nest at the tops of high cliffs, safe from the predatory attentions of skuas and Arctic foxes, but when the goslings hatch, they have to come down to feed on the grasses and sedges below. The only way down is to jump and, encouraged by their honking parents, the

MILK BAR *An African elephant calf generally feeds from its mother until she has her next baby.*

WEANING

The hooded seal is the largest of the true or earless seals. These are species living in the Arctic that can close down the openings of the auditory canals and nostrils under water, unlike the eared species, such as fur seals and sea lions, which are less highly adapted to life under water. It spends much of its time at sea and the female comes out onto ice floes to pup.

Of all the Arctic seals, the female hooded seal is the last to drop her pup, which is born at an advanced stage of development, weighing 44 lb (20 kg), very late in the pupping season. Usually the ice is beginning to break up, so she must raise her offspring quickly, ensuring that it has sufficient insulating blubber to allow it to enter the water safely before the ice floe melts and disappears.

From birth to weaning can take as little as four days – the shortest period of weaning known for any mammal. The hooded seal's near relative, the harp seal, is also quick, taking just ten days, while other Arctic seals tend towards two weeks.

Hooded seal mothers produce very rich milk with a high level of butterfat. The pup receives 3-10 pints (2-6 litres) of milk a day, so that in four days it is able to double its bodyweight from 44 to about 88 lb (20 to 40 kg).

There may be a number of reasons for late pupping and fast weaning. A short and intense period of suckling could mean that a mother loses less bodyweight, and that she would not therefore need to leave her pup alone while she goes off to feed during the short period it is dependent on her. A late birth also reduces the risk of predators such as polar bears crossing the pack ice, which by then is becoming unstable, and taking the young.

At the other extreme, the humpback whale calf is still drinking mother's milk at two years, and the African elephant calf is taking milk at three, four or even five years old. A newborn elephant calf appears slow to learn where to obtain its milk supply, first attempting to suckle from the side of its mother's leg or even from another elephant. When it does finally find the source, it is fed on demand for the next few years. The baby walks up to its mother, touches her with his trunk and the mother obliges, rarely denying her offspring access to the nipple.

tiny balls of down freefall the 500 ft (150 m) drop, either to bounce on tufts of grass and run away, or to hit the rocky scree and become an instant meal for a waiting fox.

ANIMAL NURSERIES

Some parents are unable to take their offspring with them when they go to feed. This is a particular problem for lone parents, but one way of minimising the danger is to leave the youngsters under the watchful eye of other adults.

The tiny mother lace bug, which lives on the east coast of North America, looks after both her own offspring and youngsters that have been left with her by other lace bug mothers. Why some lace bugs are happy to guard while others leave is not known, but whatever the reason a baby sitter is very energetic in defending her charges. She can only do so, however, if she keeps the group together in one place – on a leaf, for instance. If the food on one leaf runs out, or the bugs are attacked, the lace bug mother shepherds the nymphs to the next available leaf. She stands at petiole (leaf stem) junctions on the main stem, blocking off alternative routes so that all the nymphs head for the right leaf.

Eider-duck mothers also use crèches. As soon as they start to sit on the nest they are abandoned by their drakes, and they do not eat at all during the incubation period. As a result, the mothers are very hungry indeed by the time their young hatch out. The problem, however, is that their food, which consists of mussels found only in deep water, is too far from shore to take a young family, and in any case they would be vulnerable there to attacks by gulls and skuas. The answer is to gather all the youngsters together in the shallows, where they are looked after by a few female minders while the rest of the mothers go to sea to feed.

The largest crèches in the world must be those of the free-tailed bats of North and Central America. In certain caves, such as Bracken Cave, Texas, up to 20 million bats pack into caverns. When the mother bats leave the caves to feed, the naked pink babies are left hanging from the roof, packed 2000 to the square yard (m²) in order to minimise heat loss. Here they jostle about, moving up to 18 in (46 cm) from the point where their

ON THE RUN *Baby wildebeest must run with the herd within hours of being born. The slow and weak fall prey to lions, leopards, cheetahs and hyenas.*

HANGING NURSERIES *A Mexican free-tailed bat uses sound to locate her baby among the hundreds packed together for warmth.*

mothers last left them. It was once thought that the bats feed any youngster that grabs a teat first but, by capturing mothers with suckling youngsters and making genetic tests on both, researchers have established that a returning mother finds and feeds only her own baby. She flies in to the point where she last left her offspring and calls. The baby responds, and even in the noise they find each other. Having accomplished her mission, the mother leaves the crèche and roosts in a quieter part of the cave.

When the baby bats have grown up enough to look after themselves, they leave the safety of the cave. Waiting for them are bat hawks, which swoop across the cave entrance and grab inexperienced youngsters within minutes of their first flight. Of the 10 million bats born each summer in Bracken Cave, 7 million will be dead before a year has passed.

LEAVING OR STAYING AT HOME

There comes a time when baby must grow up and stand on its own two, four, six, eight or more feet. Whether it leaves home altogether or stays with the family group depends very much on the way in which the family is organised.

Most youngsters reach a time in their lives when they have to go it alone and put into practice the skills they have acquired in their early learning schools. In the natural world, the breaking of parental bonds can be quite sudden, with the adolescent simply wandering off to a life of its own. A southern right whale mother, for example, fails to retrieve a calf that has strayed from her side. But for some, the rejection is violent. Parents may eventually chase their young ones away or abandon them to fend for themselves. A grizzly bear mother, for instance, might chase her offspring up a tree and not be there when it climbs down; a growing lion cub and his brothers or male cousins may be chased away by the pride's dominant males. A mother bobcat begins to act aggressively towards her kittens by the time they have reached their first birthday, but she would have trained them well – bobcats can hunt from about seven months old. Big-eyed, white-coated harp seal pups are simply abandoned on the ice by their mothers.

Predators, well-armed with tooth and claw, have no difficulty surviving the rigours of life without their mothers, but herbivores have to find other means of survival. Young Indian blackbuck males are expelled from the main herd when they reach maturity, and gather together in bachelor herds for mutual protection. Nevertheless, they continue to contest their places in the male hierarchy and are ready to fight for the right to a harem when it is their time to rejoin the main herd, at five or six years of age.

In Kenya, kongoni males remain with their mothers for up to two years. They are tolerated by the mature males and are not expelled from the main herd immediately. For the first ten months they behave as calves, but thereafter they must show deferential behaviour to the territorial bulls. They do this by retracting the neck and making a 'quacking' sound. At two years old the youngsters are chased away by the bulls and they gather, like young male blackbuck, in bachelor groups. By four years old they will return to the main herd and challenge the territorial bulls for the right to mate.

Some large predators find life easier if they hunt in a small gang. Although young

BEAR NECESSITIES *One day soon these grizzly bear cubs will be abandoned by their mother, or chased away from home, to make a life for themselves. Right: Young male lions live a bachelor life until they are strong enough to take over a pride of their own.*

STAY-AT-HOMES *Young Seychelles brush warblers set up their own home if space is available, but will help their parents with the next brood if no territory is free.*

female cheetahs are generally solitary, young male cheetahs often form groups of two or three, known as coalitions, in which they occupy a territory and defend it rigorously from other males, sometimes killing intruders. Male cheetahs in a coalition are healthier than solitary ones, and are roughly 20 lb (10 kg) heavier on average because they are more successful at hunting cooperatively than on their own.

Lion brothers also form a bachelors' club when they are expelled from their prides. At first their unaided hunting efforts can be clumsy and unsuccessful, but they soon learn. Before long they are strong enough to defeat a pride's resident males and take it over. It is then the turn of the lionesses to do most of the hunting, while the lions defend the pride from other males and predators, including the hyena, the lion's arch rival.

HELPERS

Not all animals separate from their parents. Young Mongolian gerbils, European badgers and South American tamarins and marmosets stay at home to help their parents rear the next litter or brood – a habit they share with fledgling Australian fairy wrens

and North American scrub jays.

Florida scrub jays, which live in patchy, open scrub and compete vigorously for the better territories, were the first species of bird to give scientists an insight into helper behaviour. As researchers had observed, both parents feed their brood, but they are assisted in the task by several immature birds and young adults. Not only does this mean that the offspring are better fed, but it prevents the parents from losing weight, thereby enabling them to breed more frequently, which in turn enables them to defend their territory better. The territory holders do well from the arrangement, but what of the helpers?

The helpers are, in fact, the parent birds' offspring from previous broods. Brothers and sisters are more closely related to each other than they would be to their own offspring, because in the case of their own young the other parent is unrelated, and there is some benefit in helping to raise your own brothers and sisters – members of your own family – rather than offspring whose other parent is not a family member. In addition, young inexperienced scrub jays, up against the more experienced pairs, are unable to maintain viable breeding territories, and so, learning from experience, they do not try. Instead, they achieve greater overall reproductive success for their close relations by staying to help raise future broods of siblings.

There are other benefits, too. The helpers receive 'on-the-job' training, an experience that improves their own chances of rearing offspring successfully if they do get an opportunity to breed. They also gain a detailed knowledge of the lie of

BADGER BABY SITTERS *Young female European badgers sometimes look after the young of other females and help to protect them from predators.*

the land and where the best territories are to be found; from a relatively safe position as helpers they are therefore better able to capitalise on any territories that become vacant. And, as the parent birds age and decline, one of the helpers might well take over the family territory.

This view of why it pays to be a helper was tested in an experiment with Seychelles warblers on islands in the Indian Ocean. In 1968 the species was on the brink of extinction. Only 30 birds remained, all on the island of Cousin, but a recovery programme was successful, and by 1980 there were 400 birds occupying the entire island. Every corner was part of a warbler's territory, with the result that extra birds had to bide their time as helpers, up to ten assisting in defending a single territory, building the nest, incubating the eggs and bringing food to the nestlings.

In 1988, 20 individuals were shipped off to another island, Aride, to help build up numbers, and two years later a batch of 29 were moved to neighbouring Cousine. The vacant territories on Cousin were rapidly filled by birds that had been helpers, although those birds that had helped on good sites did not move to poor ones. The territories on Aride filled quickly and the new generation became helpers, proving that cooperative breeding of Seychelles warblers is clearly linked to the availability of suitable nesting sites and territories.

The loss of a litter can also trigger cooperative behaviour. The female European badger was always thought to bring up her own cubs and not those of any other females – that is, until the badgers of Wytham Woods in Oxfordshire were seen

to have helpers. In one study, a pregnant badger gave birth to three cubs and was helped by another mature female, thought to have lost her litter, and a one-year-old female. One of the baby sitters helped the mother to carry the cubs from their mother's burrow to her own, where they stayed with the baby sitter while the mother went foraging. Baby-sitting duties were taken very seriously, for the baby sitter succeeded in putting a fox to flight and chased away a male badger that had bitten one of the cubs.

It is not always a case of willing helpers, however. Some may be pressed into service by their tyrannical parents or, indeed, by birds from other groups. African white-throated bee-eater fathers, for example, harass their male offspring to such an extent that they abandon any hope of nesting themselves and are forced to help their parents to raise the new family. Australian

BRAVE BABY SITTER *A badger defending cubs from a marauding fox may not be their mother. Badger 'helpers' sometimes baby-sit when the mother is away foraging.*

white-winged choughs kidnap their child-minders. Baby choughs receive a specialist diet of insects so scarce that up to seven adults are needed to supply it, but there are never enough helpers to go round. In order to overcome this shortage, larger families begin harassing smaller ones, destroying their eggs and nests. They engage in exciting display battles, when the birds wave the white tips of their wings at each other, spread their tails, engorge their red eyes with blood, and call vigorously; during the confusion, fledglings from the smaller groups are coerced into joining the larger groups. Helpers at the nest only get an opportunity to breed for themselves when older and more dominant birds die and territories become vacant. Some helpers may never get the opportunity to breed at all.

Most helpers are related to both the parents and the offspring they are helping but, as the Australian choughs have illustrated, this is not always the case. Young Mongolian gerbils tend to stay with the family and help raise the next generation, but since each generation might be sired by a different father they are not necessarily closely related. There must be some other benefit for

NEST-BUILDERS *Mongolian gerbil parents are helped by young males and females, who in return learn about nest-building and raising a litter. Overleaf: Brother cheetahs hunt together after leaving their mother.*

the first generation in staying at home. A young female might gain some insight from her mother into how best to care for her young, but it is the male gerbils that gain most. Parent gerbils that have been helpers *continued on page 40*

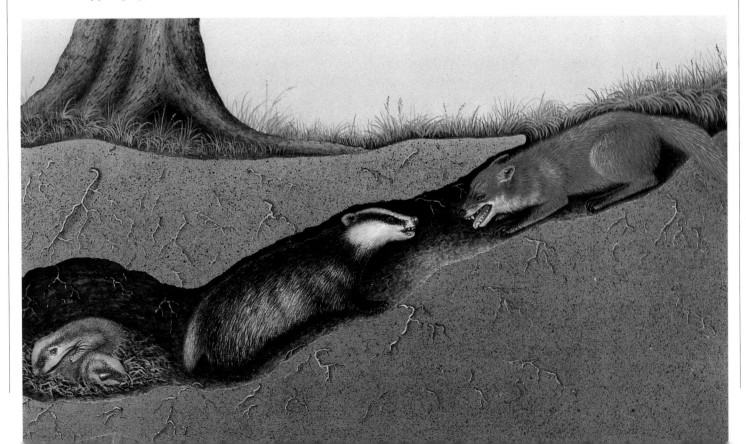

tend to have their families earlier and are more successful in rearing them than gerbils that have not had experience as helpers. The reason, it seems, is that an exhelper male is better at nest-building, and this gives his pups a better start in life.

FAMILY LIFE

Swans and geese tend to leave their summer breeding grounds in family parties, and flock to feeding sites where they spend the winter. Barnacle geese stay together until after their first winter, when the father chases the children away, giving their mother full access to the food supply so that she can fatten up before the spring migration.

White-fronted geese may stay together as a powerful extended family for up to seven years. Large families, it seems, dominate the winter feeding grounds, so it is of greater advantage to a youngster to bide his or her time with the family and remain high in the hierarchy.

Some maturing adolescents stay in their original social groups, while others move on to join neighbouring groups. Male chimpanzees tend to stay with their own group, whereas females will migrate to neighbouring communities during an adolescent oestrous (sexually receptive) period. There they are the centre of attention from the resident males, which shield them from hostile females until they are accepted by the group. Female gorillas also move out at puberty. In baboon society, however, it is the males that leave as they reach maturity. The females remain and form close bonds with other females in the troop – and it is this dense network of relatives and friends that the young male baboon has to penetrate before he is accepted by his new troop.

Whether young animals stay or go, they must adapt quickly to the sudden change in their circumstances. Some – such as insects – have in-built, inherited instructions that enable them to cope with the new demands. Others, such as birds and mammals, must learn, by copying relatives and neighbours and from the experiences of early life, how to survive without mother.

MEETING THE FAMILY *A young male chacma baboon (above) who comes a'courting has to penetrate the close-knit family circle. Below: Mute swans live in family groups, gathering in large flocks on open water where there is adequate food.*

A PLACE TO LIVE 2

PARTNERS *The zebra lets the oxpecker pick off parasites in return for lookout duties.*

THERE IS NO PLACE LIKE HOME, BUT HOME FOR AN ANIMAL IN THE WILD CAN RANGE FROM A WELL-DESIGNED AND CONSTRUCTED SHELTER IN WHICH IT CAN HIDE AND RAISE A FAMILY, TO A LOOSELY DEFINED TERRITORY IN WHICH IT CAN ROAM AND FORAGE FOR FOOD. HOME CAN BE A ROCK IN THE OCEAN, A GRAIN OF SAND IN THE DESERT, OR A TREE IN A FOREST. IT CAN BE AN OCEAN CURRENT OR A MIGRATION ROUTE ACROSS THE AFRICAN SAVANNAH. THERE ARE BURROWS, NESTS, HOLLOWS, TUNNELS, CRACKS, CREVICES AND CAVES, BUT WHATEVER PLACE AN ANIMAL CALLS 'HOME', THE OCCUPANT MUST ENSURE THAT IT IS SAFE, SECURE AND SEPARATE FROM THE HOMES OF OTHERS OF ITS KIND.

PRIVATE PROPERTY *A male mongoose marks his territory.*

FINDING A NEW HOME

The most desirable properties in the wild are often occupied and unavailable to new-home seekers. Some animals solve the problem by moving in with parents or neighbours, while others go on long journeys in search of the ideal home.

At some time in an animal's early life, it must strike out and find a niche of its own. The ways in which animals from different species leave home and disperse are determined by physical conditions, such as mountains and valleys, local climate, the distribution of potential nest sites, sleeping sites and possible sources of food, and the behaviour of animals towards others of the same species.

A social animal – one belonging to a species that lives in well-ordered herds or troops – may not need to go far to find a home when it reaches adulthood, for as often as not it is already part of a group and therefore already has a home. A killer whale or orca from one of the resident pods, or schools, patrolling the waters around Vancouver Island, Canada, stays with its family group in its own home range all its life. Similarly, a female African elephant remains with her own herd, under the watchful eye of the all-powerful matriarch. Female rhesus monkeys, from northern India and Pakistan, also stay with the family, while the males go to other troops. Each troop, therefore, may contain several female ancestral lines, with the mothers, daughters, granddaughters and even great granddaughters of one family sitting together in huddles. Most songbirds, on the other hand, are likely to compete with their parents for food, living space and future partners, so when they grow up they are not tolerated at home.

The distance an animal travels in its search for new territory will depend on several factors, including the size and shape of the area containing the resources it needs to survive, the influence of winds and ocean currents and, not least, the mobility of the animal itself.

RECOGNISING HOME

Dispersal to new homes might appear to be a problem for sessile animals – those where the adults have limited or no mobility, such as corals, barnacles and marine snails and worms. However, they overcome this by producing free-floating larvae that form part of the zooplankton, the mass of tiny creatures that drift in the ocean currents. These larvae appear to float passively with the current for a time, finally settling haphazardly on the bottom, but their method of finding a home is in fact quite deliberate. Planktonic larvae know when and where to settle,

SOCIETY GIRLS *Young female rhesus monkeys do not leave home, but must establish their rightful place in a complex, multi-generational female society.*

and have some degree of control over their movement so that they can manoeuvre into the right place at the right time.

The red abalone is a large marine snail that lives for most of its adult life on the sea-bed below the kelp forests of the California coast. At breeding time, the red abalone releases sperm and eggs and, after about 14 hours, the fertilised eggs develop into larvae, which are no more than $1/100$ in (0.254 mm) across. These larvae drift in the current, but they can also swim to the bottom in search of a suitable spot to settle. The head of each larva has a crown of long, beating hairs, called cilia, which resemble a tiny floor mop and can propel it through the water. It does not, however, settle in the first place where it lands. Rather, the larva bounces along in a behaviour known as 'bottom-hopping': it sinks to the bottom, remains there momentarily, and then propels itself back to the surface. This process is repeated time and again until the abalone larva chances upon a patch of rock-encrusting coralline red algae, from which it will obtain its first meal. The larva recognises the algae by a

FUSSY SETTLER *The free-floating larva of the lettuce coral settles on a particular species of coralline alga before it grows into the adult coral.*

BOTTOM-HOPPER *Although the adult abalone is confined to the bottom of the sea, its larvae hop over the seabed in search of a suitable site in which to settle.*

chemical on its surface that triggers the larva to settle in that particular spot and begin its development into the juvenile snail.

While abalone larvae recognise and settle on any species of coralline algae, the free-swimming larvae of lettuce corals are rather more particular. Lettuce corals' offspring find the best sites by recognising chemicals, unique to the species of algae they are seeking, in the algae's cell walls. The lettuce coral larvae prefer to settle on types of tropical coralline algae that encrust the hidden undersides of dead coral debris, where the growing lettuce coral will be less likely to be noticed by grazers such as parrot fish.

In much the same way, adult barnacles

secrete a substance in their cuticle, or hard outer layer, to encourage barnacle larvae, known as cyprids, to settle down around them and, it is thought, to cross-fertilise when the barnacles reach maturity. They achieve this with the help of the longest penis compared to body size in the animal kingdom – ranging from 3 or 4 times the body length in some species to 30 times in others.

Barnacle larvae also put down on bare rocks and pilings with no adults present, but they do not settle alone. The cyprids secrete an adhesive from their antennules, which they use to test the suitability of the surface; the more they test, the larger the amount of sticky secretion they leave. A chemical in this 'footprint' serves to attract other cyprids, and they all settle in the same place to start a new colony. The chemical remains active for about three weeks, ensuring that the site remains attractive to passing barnacle larvae throughout this vital settlement period.

The larvae of barnacles, polycheate

continued on page 46

WHALE HOPPING

Close to the deep-sea hydrothermal vents or hot springs on the bottom of the sea, there are entire communities of animals, including 10 ft (3 m) long red tube worms, giant clams and mussels, white crabs, purple octopuses, spider-like tube worms, and pink fish, living mainly on bacteria that feed on sulphur from the vents. It has been a mystery how these strange creatures move from one seabed vent system to another in order to establish new colonies; the vents, after all, are essentially oases on a sterile basaltic ocean floor. There is, however, one possible explanation.

Grey whales migrate up and down the Pacific coast of North America, travelling from their feeding sites in the Arctic to their breeding sites in Baja, California. Along the route, old animals die and their bodies sink to the bottom, where, in the inhospitable conditions of the deep sea, they rot very slowly. On one such carcass discovered by a deep-sea expedition there was a flourishing community of some of the very animals that are found at vent sites. It is possible that the larvae of these animals are swept along in the deep ocean currents, enabling the species to hop from one whale carcass to the next until they eventually reach another vent system, where they can start a new community.

worms, sea snails and corals tend to travel relatively short distances in the ocean currents, yet these species have spread all over the world. Coral larvae, for example, have a life span of about three weeks, during which they might be carried by strong ocean currents with an average speed of 25 miles (40 km) per day.

In order to travel farther than the 500 miles (800 km) or so that these currents would take them, they need assistance, and this help comes in the form of drifting flotsam, such as logs, nuts, seeds and notably volcanic pumice. The larvae settle on the flotsam and develop into adults that produce the next generation. The attached, maturing adults then breed and liberate larvae of their own, and these in their turn either freefall to suitable habitats on the seabed below, are washed up on sites suitable for colonisation, or latch onto another piece of flotsam.

Pumice is the light, air-filled rock that is expelled from erupting volcanoes. It is a common constituent of flotsam and can be found in abundance, encrusted with seaweed, barnacles, corals, tube worms, sea squirts, bryozoans, sea anemones and other sedentary animals, on the strandlines of Pacific islands.

The rim of the Pacific is particularly active with volcanoes and, in one piece of research in the Marshall Islands, it was estimated that about 100 000 coral rafts – pieces of coral that can vary in size from fine particles that are blown some distance in the wind to blocks up to 2 ft (61 cm) wide – can arrive on any one island in the group each year, some of them having

LONG-DISTANCE LARVAE
The larvae of this staghorn coral may drift hundreds of miles in search of a new home.

travelled well over 20 000 miles (32 000 km) around the Pacific Ocean before being dumped on a reef or atoll.

SOLITARY AND COMMUNAL LIVING

The way in which members of a species distribute themselves through the available area follows a pattern for each species. These patterns, known as patterns of dispersal, take two common forms, regular and clumped. Regular dispersal is seen in species where adult individuals live in clearly defined territories. Many are solitary animals. Tigers, for example, hunt and sleep alone, although they come together to mate, and the territories of males and females often overlap. Clumped dispersal is seen in social animals, those where members of a species live together in highly organised groups, which can range in size from extended families to large colonies or herds.

Regular dispersers must find and set up territories of their own when they reach adulthood. Large carnivores, for example big cats, carve out a piece of land between territories that already exist in order to claim the food sources there. They use stealth and scent-marking to appropriate new ground, carefully avoiding resident adults for fear of attack. Brother cheetahs often form an alliance to defend a territory of around 12 sq miles (30 km²). Any other male cheetahs unwise enough to enter their patch are dealt with in an uncompromising way. Researchers watching a resident trio in the Serengeti once observed three

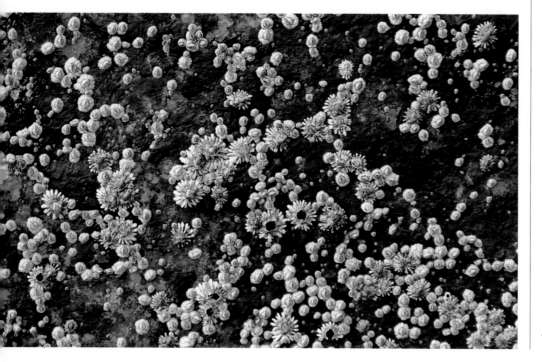

BABIES WELCOME HERE *Adult barnacles secrete a chemical that attracts young free-floating barnacle larvae to drop down and join them.*

PERSONAL SPACE *In a winter flock of black-headed gulls, each bird stands at least 1 ft (30 cm) away from its neighbour.*

nomadic male cheetahs arriving. The nomads were chased by the residents: one was caught and killed, a second was chased for nearly a mile before it escaped, and the third was badly injured but escaped.

Among the clumped dispersers – those that live in more conventional family groups, troops and herds – are many variations. There are 'contact' species, such as monkeys and apes, which are strongly attracted to each other and groom and rest huddled together – an anti-predator strategy, for the animals that huddle closest to

SUICIDAL TENDENCIES

Contrary to popular belief, lemmings do not commit suicide. At peak times in their four-year boom-or-bust population cycle, the young males forsake the congestion and food shortages, and journey together to less populated areas. They swim well and may cross rivers, lakes and fiords, even jumping off cliffs if there is no easier access to the water. If conditions are bad they may be swept into deep water or out to sea and, unable to reach land, they drown.

the centre of the group are less likely to be picked off than those sitting at the edge. By contrast, there are 'distant' species, which, although gregarious, maintain an individual distance, no matter how small. Black-headed gulls always keep at least 1 ft (30 cm) apart, and flamingos are separated by 2 ft (60 cm).

Patterns of dispersal may not be permanent, and can vary during the day or according to activity, some animals living in large groups but hunting in small ones, while others live alone but team up to hunt. Baboons, for example, come together in large groups to sleep at night, but break up into small foraging parties during daylight hours in order to cover a large area. Likewise, noisy flocks of red-winged blackbirds, numbering tens of thousands in their night-time woodland roosting sites, split up into smaller groups during the day to search for food.

The daily clocks of the scalloped hammerhead sharks work in reverse: they come together by day but hunt alone at night. During daylight hours, huge female-only congregations of hammerheads swim aimlessly up and down the underwater seamounts in the Gulf of California, moving like automatons. If the school passes through a shoal of bait fish, the sharks ignore the prey, preferring to remain in their ordered ranks, large females at the top and younger sharks at the bottom; occasionally a shark will twist and turn in a so-called 'shimmy dance' and bite the shark next door, probably to assert or challenge its place in the hierarchy.

These curious congregations, which are also seen at other hammerhead sites, such as the Galapagos Islands, are thought to be a matter of seeking safety in numbers. Even hammerhead sharks have enemies, such as other, larger sharks and killer whales, and so it makes sense to gather together and rest in the safety of the school.

Most social animals must establish their status in the community and abide by the rules of the group. A hierarchy or 'pecking order' is a simple convention, but the price an animal pays for the benefits of group living is a loss of individuality and the need to

KEEPING IN FORMATION
During the day, hammerhead females rest together in a group that reflect the school's nipping order.

conform. Otherwise, there would be mayhem at feeding time, breeding time and times when the group needs to move off in an orderly fashion, as every member scrabbled to be first.

HOUSE-HUNTING

At times, even social animals can tolerate their family or neighbours no more. An ample supply of food and several seasons of productive breeding may result in overcrowding, which presents a threat to the resident animals. Diseases may sweep through a population and previously abundant resources become scarce, and so a part of the population leaves. Every four years, for example, lemming populations swell enormously and the young males go in search of new pastures, mostly in a rather haphazard mass migration.

Bees swarm, and when they do they search for a new home in a remarkably methodical way, as one might expect from such an industrious insect. After a period of

intense brood-rearing, usually in late spring, a colony of European honeybees is ready to swarm. The nest becomes overcrowded, and the chemical messenger (pheromone) produced by the queen to control the colony and suppress the development of new queens (the only reproductive females in a colony) gradually becomes weaker. As a result, new queens are reared, and these fight for the right to take over the colony. The old queen, together with half the workers, leaves the colony and flies off to establish a new colony elsewhere.

The swarm flies a short distance at first and gathers in a great, concentrated mass, containing up to 15 000 bees, on a suitable site, such as the branch of a tree. Here, they wait and rest as several hundred scout bees set out, some flying as much as 6 miles (10 km) away, in a broad search for new premises. During their sorties in search of nesting sites, the scout bees become familiar with the surrounding area, building mental maps of the terrain. Several might return to the swarm with news of suitable

HOUSE-HUNTING *A huge swarm of honeybees rests on a fence-post while waiting for the scouts to return with news of suitable nest sites.*

FOLLOW THE QUEEN *In an overcrowded honeybee colony the old queen prepares to take a swarm of her subjects in search of a new home.*

sites – information they transmit by means of a zigzag dance, accompanied by much wing buzzing through the dense swarm. The line and tempo of the dance indicates the direction and distance of suitable new home sites.

Bees are fussy about the size and position of their new home, and a site inspection can last for over 40 minutes. They need a cavity – in a hollow tree, for example, or in a large hollow in rocks or in a cave – large enough to build sufficient honeycombs for winter storage, yet not so large that it cannot be kept warm when the weather gets colder. The entrance must be small and easy to defend, about 6 ft (1.8 m) from the ground to deter ground predators, and it must face south in the Northern Hemisphere and north in the Southern Hemisphere so that foragers can warm up on cloudy days. Entrance holes at the bottom of nests are preferred to those on top in order to minimise heat loss.

With so many scouts and such a range of suitable nest sites, the swarm must come to an agreement on which site to choose. Each scout returns with news of a possible site and performs its dance. Some bees dance more enthusiastically than others, for it is the exuberance of the performance that indicates the degree of suitability of the site. When a more restrained dancer meets an enthusiastic one, the former listens and watches, before heading off to the latter's site to conduct a personal inspection. Eventually, all the scouts will have inspected the site found by the liveliest dancer, and the entire corps de ballet of

scouts will be performing the same dance.

While the scouts are out and about, the rest of the swarm is inactive and consequently the bees' body temperature drops to such an extent that if the branch is shaken, the bees simply drop in a heap on the ground: their wing muscles are just too cold

HITCHING A RIDE

Very small animals, such as mites and spiderlings, travel vast distances to find a new home by hitching a ride. Mites travel from nest to nest on the faces of bumblebees, while a spiderling, such as a baby wolf spider, releases a long silken thread from the silk glands in its abdomen. The thread catches the wind and the spiderling floats away. The bean aphid flies upwards, attracted by the sun's ultraviolet rays, and is carried aloft by warm air currents; in this way, it is able to fly over mountains and across deserts and oceans to find new home sites many miles away. Occasionally, these tiny animals of the 'aerial plankton' are swept too far, and end their brief lives on glaciers and snow fields.

to work. Just before the flight to the new home, however, the swarm warms up, by the vibrating of wings, to a temperature at which flight muscles work best, the links between bees break down, and, one layer at a time, the swarm peels off. At first, the bees circle above the resting site in a cloud with a diameter of up to 33 ft (10 m). The scout bees fly across the swarm, indicating the direction in which it should head. With the formalities complete, the swarm is ready to

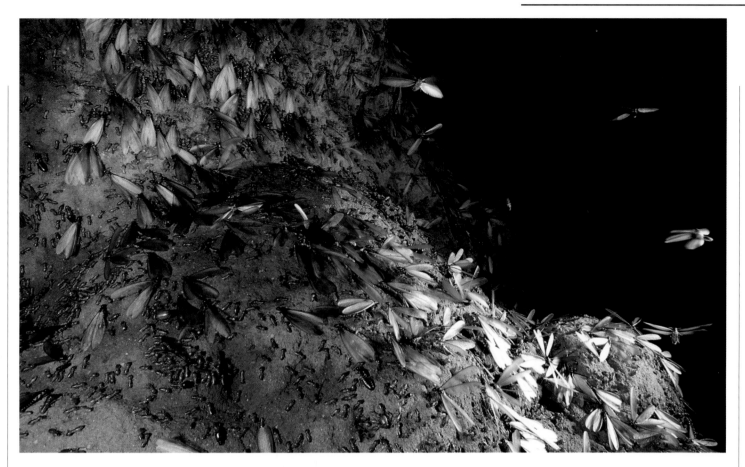

MASS EXODUS *Sexually active winged termites are guarded by soldiers and workers as they leave the mound to set up new colonies.*

move off, slowly at first but eventually streaking through the countryside at 6 mph (10 km/h).

When they reach the new site the scouts drop down and release a special pheromone that causes the rest of the swarm to gather around. In a noisy whirlwind, the swarm streams into the new nest chamber, cleans out any debris, and begins to construct new combs. The bees have a new home.

Other social insects, such as ants and termites, do not make special home-searching flights. In the case of termites, some adults develop wings at certain times of the year and take advantage of the frantic activity at mating time to disperse. Like bees, an entire colony of termites is controlled by pheromones secreted in the saliva and faeces of the queen and spread throughout the nest by mouth-to-mouth contact between workers and soldiers. Occasionally –

usually after a major rainstorm – winged termites of both sexes, capable of reproduction, fly from the confines of the termite mound to start new colonies elsewhere. The male locates a female by smell, detecting a secretion from a gland on her underside. Females from species that grow fungus within the nest for food production always carry a piece of fungus from the old nest with them in order to grow a new culture in the new home.

Whereas ants and bees have a once-only mating on the nuptial flight, a termite female and her consort become lifelong partners and will produce a whole colony together. Out in the open, termites are vulnerable to many predators, but the survivors find a suitable crack or cranny in a rock, or bore a hole in

THE QUEEN CONTROLS *The queen termite dwarfs her entourage. The odours she produces determine how many reproductives are reared to be ready when the weather is right for dispersal.*

wood or soft ground, in which to start a new colony. On reaching the ground, the termites lose their wings and the male consort, attracted by the female's scent, runs after her in a 'courtship promenade' which may last for a couple of days.

The pair first build a chamber, which will eventually become the royal chamber at the centre of the termitarium, where the female grows and develops to the point where she can produce up to 36 000 eggs. Here, queen and consort will spend the rest of their lives incarcerated in a single room in their concrete tower, maybe for as long as 20 years.

LIVING TOGETHER

Good communications between animals are the key to living together successfully, whether the animals are of the same species or not; and nature has concocted some curious and highly beneficial partnerships.

Many animals depend on others to live successfully, although the nature of the relationship varies from species to species. There are animals of different species that live together in mutual harmony, such as the hermit crab, the ragworm and the sea anemone – a behaviour known as symbiosis. There are animals that take advantage of others, often causing them harm, known as parasites. They might live on their host, like the tick, or in their host, like the tapeworm. And there are social animals, animals of the same species, whose social organisation can vary from simple cooperation between a male and a female to the complex societies of ants or baboons.

Social animals organise their groups in one of two ways: in some species there are dominant and subservient individuals, as in a troop of monkeys; others are arranged in relatively independent family groups, such as pods of killer whales. Among the former there are highly social, or eusocial, animals in which a single dominant individual, such as a queen honeybee or a queen naked mole rat, controls the entire colony.

Social animals develop effective means of communication within their groups. In a pack of wolves, for example, visual signals, sounds, touch and smell are all vital in maintaining order and status and ensuring a successful hunt. A subservient wolf will indicate its status to a more dominant member of the pack by placing its tail between its back legs and rolling over, thus exposing its neck.

The infamous wolf howl serves several purposes. Firstly it is given before going hunting, as a means of checking that every member of the pack is ready to hunt. When the hunt is over, it serves to gather the pack back together again. The communal howl also tells neighbouring packs not to trespass on the resident pack's territory. Sometimes wolves cheat. The calls bounce off trees and

HOWLING SUCCESS *A grey wolf's howl serves not only as a roll call for members of its pack but also as a 'keep out' signal to neighbouring packs.*

STAKING A CLAIM *A white-tailed deer stag daubs a territorial scent mark on a twig with a secretion from a gland in the corner of his eye.*

leg so that they can sniff each other's genitals – behaviour that is security pass and membership card rolled into one. The anal gland also produces a pungent creamy paste, known locally as 'witches' butter', which the animal rubs onto grass stems at hyena-nose height as a semi-permanent territorial scent mark – a warning to other hyena clans to keep out.

QUEEN RAT

Although the chemically controlled life of eusocial insects such as bees, ants and termites is well studied and documented, two mammals – the naked mole rat of East Africa and the Damaraland mole rat of Namibia – have the same social structure as insects.

Touch, which is concentrated in the few

hairs to remain on the rodent's body, is an important sense for the naked mole rat. Together with the tail, the hairs enable the naked mole rat to appreciate the diameter of its tunnels. However, by far the most important sense for a mole rat colony is smell. The naked mole rat queen, like her bee and ant counterparts, exercises strict control over the colony with pheromones, which she secretes in her urine and deposits in the colony's toilet, thereby spreading them to the rest of her subjects.

As with termites, the queen mole rat and her consort are the sole breeding adults. The queen lives mainly in a breeding chamber, giving birth to a dozen babies about four times a year. The rest of the colony, although not sterile, are prevented from breeding, their hormones suppressed by the smell of the queen's pheromones. There are two castes, workers and non-workers. The non-workers stay close to the queen and give the appearance of being

SURPRISING SOVEREIGN
Naked mole rat society is controlled by odours emitted by the queen, behaviour more usual in insects than mammals.

rocks as they pass through the forest, creating echoes and reverberations. In this way small packs are able to fool other packs into thinking that they are bigger than they really are.

Life in a ring-tailed lemur group is dominated by signs and smell. The members of a travelling band hold their black-and-white-striped tails in the air as a contact signal, and they scent-mark as they go, sometimes urinating on their feet to spread the message that lemurs from other groups should keep away. Gibbon pairs sing intricate duets in order to warn their neighbours to keep away and to maintain the bond between partners. Baboons flash their eyelids and groom each other constantly. And fin whales, living in scattered herds with individuals many miles apart, talk to each other using very low-frequency sounds that carry for miles across the ocean.

Smell is a particularly useful channel of communication because odours linger and the message may therefore last for some time. This is certainly true of hyena society. When spotted hyenas meet, they protrude their anal gland in greeting, and lift a hind

soldiers guarding the royal breeding chamber. The workers dig burrows, maintain the nest, and fetch food and bedding.

If the queen is killed, several of the female workers begin to develop sexually. After a short time and, unlike queen honeybees, without a fight, one female is acknowledged as dominant to the others and becomes queen. The unsuccessful queens revert to non-breeding status and become workers once more.

Workers on burrow excavation duties dig together; this is a cooperative procedure, with several workers forming a miniature production line. At the head of the line a worker chisels out the soil using its large incisor teeth, the lip-folds behind the teeth preventing earth from getting in the

WORKING ON THE CHAIN GANG
Naked mole rats dig their tunnels together, moving forwards through the soil like a conveyor belt.

mouth. The next in line gathers the loose particles of soil behind its body and shuffles backwards until it reaches the burrow entrance. Those behind straddle their legs and move forwards over the top of those moving backwards, each one becoming a soil-pusher when it reaches the head of the line. Eventually, a continuous chain is formed, those with their bellies against the floor pushing soil backwards until they reach the burrow opening and then moving forwards again. At the entrance, one individual kicks the soil out of the burrow with such force that it resembles a small volcano erupting.

For many years, the naked mole rat was considered to be the only eusocial (highly social) mammal, but in 1993 researchers discovered that its hairy cousin, the Damaraland mole rat, has a similar, but not identical, social system. The Damaraland mole rat colony has fewer individuals – on average 16 animals – including a single reproductive female, the queen, and her consort.

The queen produces about three litters a year, each containing up to six pups, while the rest of the colony maintains the sealed network of burrows, collects food and places it in a central store, keeps the babies warm, and prevents them straying too far from the breeding chambers. The males from each species differ: naked mole rat males have reduced levels of the male sex hormone testosterone, whereas the Damaraland – both the queen's consort and the non-breeding males – have much the same levels of testosterone. The ovaries of female naked mole rats are poorly developed, whereas Damaraland ovaries are functional but do not produce eggs. These major differences indicate that the social systems of the two mole rats have evolved quite separately.

AN ARMY AND ITS FOLLOWERS

The queen army ant of South and Central America has one of the greatest armies on Earth. Her subjects might total 500 000 individuals, whose lives are rigidly controlled

SOCIAL LIFE IN A WEB

Spiders are generally thought of as solitary predators, often eating their own kind, but some of them actually work together. Unlike ants and bees, social spiders do not show any division of labour, but by cooperating in building and repairing webs, and overpowering prey, they can construct much larger traps and catch considerably larger prey than a spider working alone could do.

The community spiders of southern Africa, for example, spin huge, complicated, three-dimensional webs that can stretch across a gap of 43 sq ft (4 m²) and trap winged termites, flying ants, beetles, grasshoppers, praying mantises and all manner of other smaller flying insects. One colony of *Stegodyphus* spiders was seen to construct its web across the flight path of a nest of honeybees. Once one bee was caught and had produced an alarm chemical, the rescuers all piled in and were caught as well, providing many hours of good feeding for the web's residents.

There might be several nests attached to a large web, varying in size from a thimble to a football. Upwards of 1000 spiders may live there, but they do not all come out of their silken tubes at once; the numbers to emerge appear to vary with the strength of the vibrations from the struggling prey. The larger the animal, the greater the number of spiders that are called out to subdue it.

In the spider community, females outnumber males by ten to one. Each female has a single egg sac, containing about 50 eggs, which she deposits in the nest and visits often, helping the spiderlings to escape by biting into the egg-sac wall when the time is right. The youngsters are first fed regurgitated food and then

COME INTO OUR PARLOUR
Some spiders live together on communal webs and share their prey.

portions of prey. During the breeding period, the mother spiders become progressively weaker and die; a single generation of spiders lives for about a year.

With its accumulations of old egg sacs, moulted skins, prey carcasses and vegetable debris, a 'spider city' attracts the unwelcome and the undesirable for its rich pickings and good hiding places. The pyralid moth *Loryma* moves gently over the spiders' web, searching for a good place to lay her eggs. If she should encounter a resident spider, she remains motionless until it has passed by. The moth larvae, protected from their hosts by their own silken tubes, scavenge on the community's garbage. Here, they are joined by scavenging bark lice, which in turn are prey to armoured spiders. As a communal web grows, more and more visitors move in: parasitic wasps lay their eggs on the moth larvae; huntsman spiders, jumping spiders and cockroaches hide in unused portions of the web; and birds steal spider silk for nests.

There are thieves, too. Tiny *Argyrodes* spiders, resembling tiny silver bubbles, and hairy *Portia* jumping spiders regularly steal the residents' prey. But the most brazen robber must be *Archaeodictyna,* a small spider that closely resembles

the communal ones, and whose ancestors were themselves communal web spinners. This ancestry, bringing

SPIDER COMMUNE
Social spiders weave a communal web to catch large prey, but each spider has its individual nest.

with it the innate ability to step light-footed across the communal web, is thought to be the clue to the interloper's success. *Archaeodictyna* lives alongside the resident spiders, but if it arrives first at trapped prey, it steps back and waits for its hosts to kill it; only then will it move in to feast. If all the places are filled, however, the cheeky stranger crawls over the feeding spiders, tapping them with its forelegs until one moves over and makes a space for it to feed.

by their queen, and whose movements are regimented much like a human army. The troops, however, are only effective if they work together; to do so, the ants communicate with smells and vibrations. Together they form a deadly 'super-organism', for army ants are voracious predators.

The army ant's food generally consists of hard-bodied creatures, such as scorpions, spiders, tree termites and other insects, and the colony must be constantly on the move in search of new supplies. Movements are not constant, a nomadic phase alternating with a sedentary phase. After a couple of weeks of travelling through the forest, the colony comes to a halt and sets up a semi-permanent camp, where it stays for about 20 days while the queen lays her latest batch of 50 000-100 000 eggs. In the meantime, the workers set out to search the surrounding forest for food. They forage in a fixed pattern, radiating from the camp each day like the spokes of a wheel. On day one the columns head off in a particular direction, and on day two they take a compass heading exactly 123 degrees in another direction – with the result that they do not comb the same sector of forest twice.

When the eggs hatch out and there is a new army of mouths to feed, the colony returns to a nomadic way of life. At dawn, columns of workers, guarded at intervals by large-jawed soldiers, set out from the daily bivouac and go hunting. Maintaining a constant compass heading, and avoiding the columns of other colonies, which they de-

tect by smell, they travel at a speed of about 45 ft (15 m) per hour. They complete a distance of 600 ft (200 m) each day, during which they might gather up to 30 000 items of food.

Scuttling to and from the bivouac, where the queen and her entourage spend the day, there might be over 200 000 ants in a column. Side-streams split off from the main column and spread out to form a raiding front about 60 ft (20 m) wide. Not even streams are a barrier: the workers simply interlock legs and form a living bridge.

Each evening the queen's bivouac unravels like a ball of wool and, under cover of darkness, progresses down the route of the day's raiding column. Travelling for about eight hours, the queen and her retinue pitch camp again before dawn breaks, about 270 ft (80 m) down the trail.

Army ants, like most armies, also have their followers. A column of raiders is accompanied by an entire menagerie of animals, each intent on taking advantage of the mayhem caused by the advancing ants. Antbirds are the most obvious. They do not eat the ants themselves but prey upon the insects flushed out ahead of the columns.

The dominant antbirds station themselves at the head of the column, where most insects are driven out, while the lesser-ranking birds take up position behind and alongside the marching columns. In the trees above, large numbers of birds follow the procession. They are not interested in the ants, but are taking advantage of the sentinel services and alarm calls of the antbirds, which are ever alert for predators, such as forest hawks.

One antbird cheats, however, in order to steal the choicest morsels driven out by the ants. If a grasshopper jumps from the grass and a rival bird is in pursuit, it screams an alarm

ANT LEGIONS *Large-jawed soldiers protect the constantly moving columns of army ant workers as they scour the forest for food for the colony.*

WASH AND BRUSH-UP *A female kudu tolerates the free-riding oxpeckers because the birds pick off their host's cargo of ticks and mites.*

call. The competing bird is startled and takes cover, preferring to lose its meal rather than its life, and the cheat scoops up the prey.

The parade does not end there. Following the antbirds are ant butterflies – ithomine butterflies that seek out the minerals in the droppings of antbirds and locate the army ant columns by detecting odours given off by the ants. The butterflies are protected from the antbirds by a poison that they acquired as caterpillars from the *Solanum* plant, on which they fed and grew, retaining the chemicals when they metamorphosed into adult butterflies.

ANIMAL BUDDIES

Like the parade of army ants, antbirds and ant butterflies, different species of animals sometimes find it convenient to live together. In this type of relationship, known as symbiosis, all members benefit and therefore trust each other. On the African savannah, for example, several species of birds and mammals have formed an alliance. Cattle egrets often accompany African buffalo and elephants, pouncing on the many insects driven out of the vegetation by the trampling of the large mammals. Likewise, oxpeckers divest large African herbivores, such as zebra, buffalo and rhino, of their burden of ticks and mites. Carmine bee-eaters are not attracted by the prospect of juicy parasites, but simply stand on the backs of Arabian bustards to gain a better view of the insects these large birds flush from the grass. In return, the vigilant birds warn their hosts, whether bustard or buffalo, of approaching danger.

On coral reefs in tropical waters, clown fish and sea anemones help each other. The fish live among the deadly tentacles of the anemone without fear, for they are somehow protected from the stinging cells. When the small, colourful fish first approach the anemone, they appear to have

AGRICULTURAL ANTS

In the 100 million years that ants have been on Earth they have mastered the art of cooperation and living together and, in doing so, have come to dominate the aggressively competitive world in which they live. Ants build cities, roads and bridges. They weave nests, take slaves, form huge nomadic armies, herd livestock, tend fungus gardens, collect crops, store food and use tools. However, one of the most intriguing aspects of ant life is the way in which they form mutually beneficial relationships with other living things.

The classic ant relationship is the one with aphids, described by Linnaeus in 1776 as 'ant cattle'. Aphids feed by pushing their piercing mouthparts into the phloem of plants and sucking out the sap, which is high in sugar but low in nitrogen. Excess sugar is exuded as honeydew, a sugary globule that is collected by ants. As well as harvesting the sugar, however, the ants give something in return: they look after their aphid stock, affording them protection from ladybirds and spiders.

Many ant species balance their diet with other foods, which they forage from the surrounding countryside, but some rely on honeydew alone.

The ant *Dolichoderus*, from the rain forests of South-east Asia, feeds exclusively on honeydew from mealy bugs, relatives of aphids. Mealy bugs also suck sap, but their ant protectors ensure that the honeydew they produce is rich in amino acids – the building blocks of proteins, which make for strong ant bodies – by shepherding their charges towards young leaves, which contain more amino acids than older ones. When the level of amino acids in the honeydew drops, the 'flock' is moved on by their nomadic pastoralists.

Another tropical Asian species takes the farming analogy further by building stables for its insect herds.

Polyrhachis ants, using silk secreted by their grubs, glue bamboo leaves together to build shelters for their charges, and *Crematogaster* ants excavate chambers in the stems of *Macaranga* trees to shelter both their own brood and their livestock. Some *Crematogaster* ants have turned to piracy to supplement their diet, ambushing the workers of other ant species along their foraging trails and stealing their seeds.

A North American bug – the ironweed treehopper – has capitalised on its relationship with ants. The female is a good mother, with a mighty kick, protecting her offspring from predators for about 32 days,

LIVING STORAGE JARS *In order to have food available during lean times, honeypot ants use the bloated abdomens of living worker ants to store honeydew.*

until they are ready to stand on their own six feet and fend for themselves. A mother hopper attended by ants, however, can relinquish her maternal duties to ant guards, cut the time with her offspring to six days, and go and have another brood.

Leafcutter ants are arable farmers. The leaf sections removed from the surrounding forest are heaped into chambers, where a fungus is grown on the rotting vegetation. The ants feed on the fungus from these cultivated fungus gardens, making them less dependent on the vagaries of the weather. Australian honeypot ants also have an insurance policy, not for rainy days but for periods of drought. They force-feed honeydew, obtained from sap-sucking bugs, to a band of conscripts whose abdomens, filling with honeydew, swell up to an enormous size. These living storage jars, hanging in rows from the ceilings of special storage chambers, regurgitate food to the other colony members when food is scarce.

---

no immunity, but as they touch the anemone's tentacles time and time again they build up their protection, covering themselves with a mucous coat. The relationship is of mutual advantage, for the fish keep the anemone free of scraps of food and other debris, while the tentacles of the anemone protect the fish.

Clown fish society is intriguing for another reason: the fish are able to change sex. Normally a sea anemone has several fish living among its tentacles: one large female, a smaller male, and several juveniles. If the dominant female should die, the male changes sex into a female, and one of the juveniles becomes a fully functional, mature male, as the group continues to flourish in its deadly home.

A small shrimp-like creature living in Antarctic waters has gone one better than the symbiotic clownfish: it carries its protection around with it. The crustacean in question is the *Hyperiella*, a member of the order Amphipoda, which has struck up an unusual relationship with the sea butterfly or pteropod *Clione*, a mollusc without a shell. *Clione* does not need a shell, however, for it is protected by noxious chemicals that cause any predatory fish that catches it to spit it out again. *Hyperiella*, on the other hand, has no protection of its own, and is the number one prey of the fish *Pagothenia*. So, if the *Hyperiella* should chance upon a *Clione*, it captures it and carries it around on its back as a chemical defence. Predatory *Pagothenia* fish avoid any amphipods

carrying sea butterflies, and so the crafty *Hyperiella* survives.

Another relationship involves tropical cleaner fish. These function in a way not dissimilar from African oxpeckers, but they differ in having cleaning stations at permanent sites. The fish advertise their business with a blue stripe down the body. When small fish the size of gobies, or large ones such as groupers, sharks and manta rays, recognise this distincitve business sign, they queue up for attention. The cleaners first perform a little dance, indicating that they are open for business, and then the customers hang motionless in the water, allowing the fish, and sometimes cleaner shrimps, too, to clamber all over their body, into the mouth and around the gills. Here

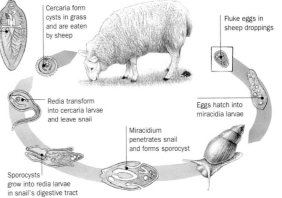

Liver fluke

Cercaria form cysts in grass and are eaten by sheep

Fluke eggs in sheep droppings

Redia transform into cercaria larvae and leave snail

Eggs hatch into miracidia larvae

Miracidium penetrates snail and forms sporocyst

Sporocysts grow into redia larvae in snail's digestive tract

they remove dead scales, skin and parasites.

The clue as to how this relationship developed can be found in the modern cleaner fish's stomach. Surprisingly, parasites form only about one-fifth of the cleaner fish's diet; they are more interested in scales and mucus from the customer, and only take a few parasites while they are feeding. Originally, cleaner fish probably took small chunks out of other fish, but because they also took away the occasional external parasite, the host began to stay put rather than flee, and the trusting relationship began to develop.

One small fish, however, mimics the cleaner fish and fools its customers. With its blue and white pattern, the false cleaner, a member of the blenny family, looks remarkably like a cleaner fish. When the customers line up, the blenny is allowed to approach

SWEENEY TODD   This blenny, disguised in the striped livery of a cleaner wrasse, will take a bite out of a larger fish, rather than remove parasites.

FLUKE CYCLE   *One species of parasitic liver fluke has two hosts, a sheep and a snail, and cannot survive if either is missing from the life cycle.*

as close as it likes, but instead of a quiet nibble at the parasites and dead skin, it uses its underslung mouth and long teeth to take a chunk of flesh, fin edge or scale. The customer, having been duped in this way, is very wary of visiting that particular cleaning station again.

INFLUENTIAL PARASITES

Whereas symbiosis benefits all parties, in the case of parasitism it is only the parasite that benefits, feeding, living and breeding on or in its host. It thus presents a threat to the host's health and ultimately to its life. Nevertheless, parasites are among the most extraordinary creatures in nature. A parasite's entire world is provided by another creature, or series of creatures, and every stage of its life cycle is dependent upon the

unwitting and unwilling host's behaviour.

The parasite can occasionally even influence its host's behaviour. There is one species of liver fluke that divides its time between a sheep, an ant and a snail. First, it is defecated by the sheep and, with the droppings, falls to the ground, where it is eaten by a snail. Inside the snail, the parasite takes the form of a small ball, coated in slime, which is ejected from the snail's breathing hole and eaten inadvertently by ants attracted to the slime. Inside the ant, the parasite creates havoc, causing the ant to forsake its nest mates and climb to the top of the nearest grass stalk. Here, it waits to be eaten by a grazing sheep, and the cycle starts all over again.

Another species has just two phases: one inside a European land snail as a liver fluke, and the other inside a bird, such as a thrush or flycatcher, as a gut parasite. Getting from the bird to the snail – via the droppings – is the easy bit. But, in order to move from the snail to the bird, the parasite has to grow inside the striped sacs which migrate, each morning, up from the snail's liver to its extendible tentacles. The normally thin tentacles become thick and club-shaped, and the skin is stretched so taut that the brightly coloured parasite can be clearly seen inside. To add to its attractiveness, it pulsates and, to all the world, resembles a wriggling, writhing larva that is highly attractive to caterpillar-eating birds. The birds peck at the tentacles, swallow the parasite, and the cycle starts all over again.

PARTNERS   Cattle egrets (overleaf) wait for the buffalo to move and disturb insects in the grass. The birds act as look-outs for the resting buffalo.

HOUSING MATTERS

While some animals build their homes from what is available,

others manufacture materials with extraordinary qualities:

silk with the strength of steel, dried mud with the consistency

of concrete, wax and paper that support complex structures.

For many animals, a patch of ground on which to feed, a convenient roosting site in which to sleep, and somewhere suitable – a forest clearing or a scrape in the ground – on which to give birth or deposit eggs is just not enough; they need some protection from the elements and from predators for themselves, and a safe and comfortable place in which to give birth and raise their young.

While some animals make themselves at home in existing features such as caves or hollow tree trunks, others must borrow, burrow or build a suitable property, and employ a diverse range of adaptations and skills in order to do this; some mix concrete, others weave with twigs and grasses, stitch leaves together, or produce silk, wax or a paper-like substance. Burrowers tend to have specialised digging limbs in order to excavate their tunnel complexes. The mole, for example, shares with the mole cricket a pair of spatula-like forelimbs that can shift large quantities of soil, while naked mole rats have long incisor teeth for digging out their tunnels.

SQUATTERS

Some animals leave the burrowing to others. The homes vacated by other animals are sound investments – after all, the original owner has already carried out the hard work – which is why puffins nesting on islands off the south-west coast of Wales, and Magellanic penguins in the Falklands, move into vacated rabbit burrows; tuatara lizards in New Zealand take over the burrows of fairy prions and shearwaters; North American burrowing owls lodge in the holes scraped out by gopher tortoises; and as many as 60 wrens may 'huddle' together in a single squirrel drey.

The original residents do not always move out when the new tenants move in. Termite mounds, for instance, are popular sites – Australia's gold-shouldered parrots make their nests in the high-rise mounds of compass termites, as do lace monitor lizards. Similarly, Mexican orange-fronted parakeets nest almost exclusively in

BASEMENT SQUATTERS
North American burrowing owls stand guard over their nest – a hole in the ground excavated by a gopher tortoise.

NEW TENANTS *Egyptian geese move into abandoned nests constructed by Africa's hammerheaded storks.*

the mounds of *Nasutitermes* termites. The parakeets benefit from the protection of the concrete-hard termitarium walls and from the resistance put up against predators by the soldier termites during an attack. The 'houseproud' termites also keep the parakeets' nest meticulously clean, removing bird-nest lice and their larvae, as well as the chicks' faeces. What the termites receive in return is not clear. It could be that their squatters provide valuable salts and other compounds in their droppings, or that the termites steal nest materials, which they find nutritious, from the birds. It seems certain that the soldier termites would soon evict the birds if they provided nothing in return for their safe haven.

Very large nests, like those of the African hammerkop or hammerheaded stork, may have an entire menagerie living downstairs, in the attic, or in the walls. A new nest, with a large roof and 10 in (4 cm) thick walls, may be 3 ft (1 m) wide and stand 5 ft (1.5 m) tall. It is plastered with mud on the bottom to keep out rats and snakes, and is covered with a thatch of plant debris, snake skins, feathers, and bits of dead animals. The nests are so solid – the giant edifice will withstand the weight of a person – that many creatures move in to cohabit. Pel's fishing owls and milky eagle owls may invade the upper dome; several species of bulbuls and wattle-eyes nest among the sticks at the side; and weavers and penduline tits hang their own covered nests from protruding sticks.

Unfortunately for the industrious hammerkops, some species covet the nesting chamber and take over completely. Barn owls, Dickinson's kestrels and Verreaux's eagles often oust the residents, while Egyptian and African pygmy geese will take over a vacated nest. Nevertheless, all these interlopers may be ousted by colonies of a very aggressive home-hunter, the African honeybee.

HOLES IN TREES AND ROCKS

Some animals seeking a roof over their head simply make do with what is readily available; fringe-lipped bats, for example, returning from a frog-eating foray in South America's tropical forests, take over a hollow tree-trunk, returning to the same tree each night. Similarly, sugar gliders retire for the day into communal nests in tree hollows in Australian woodlands, treecreepers build their nests in tree cavities, and tiny wrens pack into holes on tree-trunks, where they roost together with up to 60 birds crammed into a single shelter. But for some animals, the tree cavity is just a starting point. Like do-it-yourself enthusiasts, they tinker with their homes. The red-cockaded woodpecker of the south-eastern USA, for instance, sculpts its nest cavities in living pine trees. The damaged part of the tree produces copious resin, which flows down the trunk and gums up the scales on the belly of tree-climbing rat-snakes, preventing them from reaching the nest. Some woodpeckers even strip the bark from the trunk below the nest hole to produce an even smoother surface over which the resin can flow, making it an even more difficult fortress to penetrate. Male hornbills make

SAFE FROM SNAKES *The red-cockaded woodpecker drills a nest hole in a pine, releasing a stream of resin that prevents snakes from climbing to the nest.*

their tree-hollow nests even more fortress-like against the predatory attentions of tree snakes by sealing up the entrance with mud. The female is incarcerated in the chamber for the entire breeding season, and only emerges when the chicks are ready to fledge.

Caves, with their even temperature day and night, summer and winter, make good homes. Bats and swiftlets (cave-dwelling swifts) share the cathedral-like caves of Southeast Asia, such as those at Mulu, Sarawak. The bats roost in the deepest parts of the cave, hanging from *continued on page 64*

SPIT IN THE DARK *Cave swiftlets in South-east Asia use saliva to build cup-shaped nests that are firmly fixed to the cave walls and roof. Overleaf: Grey meerkats from southern Africa can dive down their communal underground burrows if danger threatens.*

BAT BIVOUAC *Left: Tent-making bats cut through a leaf so that it folds into a tent. Above: Costa Rican bats are almost invisible inside their tent.*

the roof by their toes. The swiftlets, which build the cup-shaped nests of saliva so prized by Oriental gourmets for bird's-nest soup, nest precariously on the walls and ceilings of more accessible caverns towards the cave entrances. At dawn their paths cross, as

SPIDER SILK

The silk secreted from the spinnerets of spiders is only 1 micron, or $^1/1000$ mm, thick – yet it has twice the relative strength of steel, and can stretch twice as far as the silk from silkworms. The silk is a protein, made up of amino-acid building blocks put together in a particular fashion: there are tough crystalline parts, which have good resistance to impact, linked together with very rubbery elastic parts, which have plenty of 'give'. Scientists are now trying to copy the chemical formula and manufacture a similar material. In the early 18th century, a French inventor did, in fact, make some stockings and gloves from spider egg-sac silk. However, as it was estimated that it would take 663 522 spiders – compared with about 350 silkworms – to make 1 lb (450 g) of silk, the research was never pursued.

millions of bats return from hunting insects during the night, and equal numbers of birds emerge to hunt for insects by day. At dusk the roles reverse – but waiting for the change of shift are the bat hawks, which swoop in and scatter the swirling clouds of bats and birds, grabbing their victims in midair.

Elsewhere in caves are smaller creatures that have modified their underground homes to suit their purposes. The curious trellis worm, the larva of a fungus gnat, spins its disorderly web between stalagmites, and moves about on a layer of slime searching for insects that have blundered into the sticky threads. In subterranean streams are the hiding places of creatures that seem to have survived from a bygone age: blind shrimps, salamanders and fish.

Underwater caves are popular places to live or hide, too, and are favoured by the coelacanth, the living fossil found in deep waters off the Comoro Islands in the Indian Ocean. It rests during the day in submarine caves, safe from the predatory attentions of sharks; by withdrawing into a cave, away from the water currents, the coelacanth also saves energy. These primitive fish are active hunters at night. Each morning the smaller ones retreat to the same cave, but the larger fish may use five or six caves within their home range, the caves extending 6-10 ft (1.8-3 m) into the lava rock of the submarine slope, at depths of 590-820 ft (180-250 m) below sea level. Some fish hide alone, but as many as ten coelacanths may be found in a single cave, each one hanging motionless in the water.

CONSTRUCTION WORKERS

The largest bird's nest is probably that of the scrub fowls from Australia and Southeast Asia, which build a nest mound over 36 ft (11 m) long and 16 ft (5 m) high. The

nest is constructed by the male. He digs a hole 16 ft (5 m) in diameter and 3 ft (1 m) deep and lines it with vegetation. Once the pit has been moistened by rain, he fills it in with sandy soil, until the mound may contain 295 tons (300 000 kg) of material.

Some animals are quite ingenious in making the best use of building materials. European rooks, for instance, crisscross the twigs in their nests in such a way as to make it difficult for neighbours to steal them. Weaver birds knit grass stems into nests, producing simple, ball-shaped nests for single pairs of birds, or great communal nests that are so heavy they can break the boughs of large trees. The southern Asian tailor bird is even more enterprising: it uses spiders' silk, shredded bark and cotton fibres to stitch banana leaves together when constructing its grass-filled nest.

Tent-making bats use leaf-bending techniques to make their tent-shaped daytime roosts. They chew through the veins on either side of the midrib, and the leaf collapses into the shape of a roof or tent. In Costa Rica, one of these tropical tent-makers is *Ectophylla*, a tiny bat no heavier than a small coin, which makes its tent from the broad, 3 ft (1 m) long leaves of *Heliconia*

continued on page 67

ADOPTED HOMES AND ACQUIRED PROTECTION

Hermit crabs the world over, whether in a rock pool, in the deep sea or on land, live in borrowed homes. A marine hermit crab will usually seek out the shell of a mollusc, such as a whelk, while a terrestrial crab suffering a severe housing shortage might end up in a discarded bottle.

Land hermit crabs find their new homes by smell, and are attracted to shells containing recently dead crabs, presumably because the shell worn by a crab of the same species is more likely to fit. When the scent of a suitable home is detected, all the crabs in the vicinity go into a home-swapping frenzy. The largest crab inspects the vacant shell, touching it with its claws and walking legs, and the rest pile into a kind of rugby scrum behind it, the largest at the front and the smallest at the back. As soon as the crab at the head of the queue moves into its new home, the next in line takes over its vacated shell and so on down the line, until they have all moved up the housing ladder.

Several species of marine hermit crabs gain additional protection for themselves by attaching one or more sea anemones or sponges to the shell; some even go so far as to take their own anemones with them when they swap shells. The crab gently rubs the anemone until it releases its hold on the old shell – which the crab can do safely because it has a hard exoskeleton that the sea anemone's barbs cannot penetrate – allowing the crab to transfer the anemone to its new shell. Some anemones even activate their own move: they bend over, latch onto the new shell with their tentacles, and then haul their bodies over in a somersault to complete the manoeuvre. When the move is completed, however, the crab may have a problem if the weight of the anemones has altered the balance of the shell. Hermit crabs have a solution: they position the anemones very carefully to take account of weight distribution. In experiments in the aquarium, small fishing weights have been attached to a shell, and the crab is seen to adjust the position of its anemones accordingly.

The sea anemones, their tentacles filled with row upon row of deadly stinging cells, provide a formidable defence for the crab. The crab gains protection from predators such as octopuses, which can rip unprotected crabs from their shells, or box crabs, which can snip their way in, by wearing a cloak of sea anemones. But the anemones gain, too. They are carried to areas where the ocean currents bring them food, and they might also share in the feast as crabs pull apart their scavenged food. They acquire something else as well – protection for themselves. Sedentary sea anemones are easy prey for predators such as starfish, but a hermit crab, taking its sea anemone with it, will simply walk away from any predator that is approaching. One species of sea anemone on the Florida coast is attacked by fireworms, but individuals attached to hermit crabs are safe because the crabs catch and eat the worms.

Deep-sea hermit crabs can be found $1/2$ mile (800 m) below the surface, on lava flows around the Hawaiian islands. They have long, spindly legs and a golden shell that is often crowned by an orange sea anemone. The 'shell', however, is not made by a deep-sea mollusc, but is manufactured by the sea anemone and consists of a very thin, parchment-like 'shell' made of chitin, the same substance that crab exoskeletons and insect cuticles are made from. The anemone builds its replica by finding a young hermit crab in a small mollusc shell and attaching itself, coating the real shell with chitin, and adding more material to accommodate the hermit crab as it grows. Following the shape of the crab's corkscrew-like abdomen, the chitin 'shell' eventually resembles a real shell.

This arrangement is to the anemone's advantage as the hermit crab can carry it across the deep-sea bed where, in the muddy or sandy conditions, the sea anemone would otherwise have nowhere to attach itself. The hermit crab gains by always having the right-sized shell without the trouble of abandoning its current shell and searching for a larger one. After all, there are not too many large snail shells to be found in the deep sea; and furthermore, the shells of marine molluscs tend to be made of calcium carbonate or chalk, a substance that dissolves easily in the ocean depths. In this environment, the fake shell manufactured by a sea anemone wears better than the real thing.

Travelling Companions
The sea anemone and the hermit crab may stay together for life, even though the crab changes its shell regularly.

plants. For most of the year, up to 17 individuals of both sexes huddle in one or other of these tents, which is usually at a convenient flight height of between 6 and 10 ft (1.8-3 m) off the ground. But come the breeding season in April, all the pregnant females gather in a 'maternity' tent, guarded by a single male. The rest of the males occupy bachelor tents until all the youngsters have grown up and are independent of their mothers. At this point, the sexes mix again and return each dawn to their more usual unsegregated roost sites. The pale, almost white, colour of *Ectophylla* bats is thought to help camouflage them while under their tents; the green light, filtering through the leaves, reflects off the bats' fur, making them almost invisible.

The most enterprising tent-maker, however, is probably Peter's tent-making bat from Panama, which creates more of a tepee than a tent, in the Coccoloba tree. Like other tent-makers, the bat nibbles at the midrib of leaves, but it does not cause the entire leaf to collapse, for it restricts its gnawing to the midrib farthest from the trunk, causing that part only to bend. Then, working in an upward spiral, the bat (or bats – for no one is sure if one or several bats construct the tent) modifies the leaves above, biting into the midrib gradually closer to the trunk as it goes. The result is a rainproof, tepee-shaped tent, with the upper leaves drooped over the lower leaves, like slates on a roof.

INSECT BUILDERS
The ultimate builders must be the social insects: wasps make nests of paper; honeybees build homes of wax; ants construct chambered colonies underground; and termites erect major architectural towers, complete with automatic air conditioning, made of a

HANGING PAPER *Social wasp workers (left) in Mexico regurgitate chewed wood and saliva to construct elaborate nests made of paper. Right: The towering nests of the Macrotermes termite can reach heights of 26 ft (8 m).*

mud and mucus mixture that hardens to the consistency of concrete.

The shapes and sizes of termite mounds vary from one part of the world to another,

---

POLYESTER BEES

Humans are not alone in producing polyester, which is a type of lightweight resin. The female *Colletes* bee digs burrows in the ground, often in suburban lawns, in the eastern USA. The burrow has ten side tunnels, each ending in a brood chamber containing a dollop of nectar and pollen and an egg. The chamber is lined with polyester secreted from a gland on the mother bee's abdomen and brushed on with her tongue. The transparent liquid hardens to form a tough coating to the brood cell, much like a plastic bag.

---

and often depend on local weather conditions. The termites of the African savannah construct high towers, like chimneys, to dissipate the heat generated by the colony, while the termites of the tropical rain forest floor construct mushroom-shaped mounds, the roofs of which function like umbrellas to protect the colony below from the daily downpour. But the most extraordinary termite mound is to be found in Australia. The 6 ft (1.8 m) high mounds of the compass

termite stand like rows of modern high-rise office blocks. About 2 million termites live inside what, in equivalent human terms, would be a skyscraper some 5 miles (8 km) high.

The mounds of compass termites, which are usually found in clusters, all face in the same direction: the narrow sides point north and south, and the broad sides face east and west. At midday, when the sun is at its hottest, it is in the sky directly to the north and the smallest surface area of each mound is exposed. In the cool of early morning and late evening, the broad faces absorb the maximum amount of solar energy to warm the colony.

For a long time researchers wondered how the compass termites know which way to build, although they were sure that termites can detect the lines of force in the Earth's magnetic field. The queen termite, as often as not, lies in her chamber with her long body axis running either north-south or east-west, depending on the species. In the laboratory, however, termite queens settle down at random if the geomagnetic field is artificially neutralised with the use of large magnets. But magnetism appears to be only part of the story.

In field experiments in Australia's Northern Territories, researchers placed *continued on page 70*

TROPICAL CAVE LIFE

Caves are comfortable and convenient places to live, for the microclimate varies little throughout the year, day and night, wet season or drought. Living in the darkness of many of the world's caves, there is often an extraordinary collection of animals, many yet to be recognised and recorded by science. In the great cathedral-like caves of Malaysia there are crabs that behave like spiders; snakes that squawk and catch bats in midair; dinner-plate-sized toads, blind amphipods (shrimp-like crustaceans living in the subterranean streams) and worms; and shrews that squeak at frequencies much higher than we can hear.

The resident animals are able to survive there because the bats and swiftlets, which nest or roost in the caves, bring in food from the surrounding forest. In addition, their droppings and the other debris, including dead or dying animals, that rains down from nests and roosts in the roofs of the caves keeps the entire cave community going.

The mountains of bat guano on the cave floor are home to a myriad of scavenging cockroaches, which sift the waste for fragments of food, but the most spectacular cave insect is the cave cricket, a voracious predator. It has great muscular legs and huge jaws, and is well able to climb up the cave walls to the swiftlet nests, where it can tear a swiftlet chick apart or crush an egg.

Cave racer snakes follow well-worn trackways, like ribbons of polished mud, into the swiftlet caves. Here, they find constrictions in the passageways, coil their tails around a convenient stalactite, and hang into the void with open mouths. In the confusion of the inevitable traffic jam, as bats and birds slow down to negotiate the narrow passage, the snakes grab their victims in midair. It is thought that the snakes detect air-pressure waves from the flapping wings.

The detection of pressure waves is common in cave animals, such as the blind hunting spider of the Malaysian caves, which finds its way about by running on three pairs of legs, like an insect, and waving its elongated forelegs, which are covered with sensory hairs, ahead of it, much like antennae. In this way, it finds small cave crickets, its usual prey.

If the stream on the floor of a cave floods, crabs emerge to feed on the water-logged guano or on baby swiftlets that have failed to complete their first flight and ditched on the cave floor. The crabs fight for each prize, ripping the chicks apart between them. In the darkest recesses of the caves is another crab, which has lost all pigment, has feeble eyes and very long legs. It scuttles about, both in and out of the water, and is able to climb the cave walls much like an amphibious spider. Catfish, well adapted for life in the dark with sensory barbules on the chin, also live in the water, feeding on cave shrimps.

Strange shrimp-like isopods and amphipods are closely related to a species found only around the Mediterranean. It is possible that these species have been living in ground waters for so long that they were in existence before the continents started to drift, and that the Malaysian caves are a time capsule that has trapped animals from a previous age.

CAVE CREATURES *All manner of strange creatures hide in the dark of tropical caves in South-east Asia. Bats roost and cave swiftlets nest in the roof, while cave racer snakes hang from stalactites and snatch their prey in midair (top centre). Long-legged cave crickets (right), cave spiders (bottom right) and multi-legged poisonous centipedes (left) scuttle across the walls and floor in search of prey, while cockroaches (bottom right) and freshwater crabs (bottom left) scavenge on the guano and dead bodies which fall from above.*

large magnets around some young compass termite nests, and unmagnetised iron bars, as controls, around others. The magnets were placed in such a way that the termites, if using magnetism, would build their nests with the narrow axis running east to west. Several months later, the researchers found that the control nests had grown normally but that the nests surrounded by magnets had been abandoned. It is thought that the colonies had been able to detect the magnetic fields but had received conflicting information from another orientational clue, probably the daily movement of the sun. They had been completely confused and had simply given up building the nest.

Compass termites are not the only insects to orientate their homes in a particular direction. In a colony of wild honeybees, the stack of honeycombs is orientated in a

HIGH-RISE TERMITES *These flat-sided slabs of 'concrete' are shaped to take advantage of the sun without overheating.*

FORTRESS ANT *Mulga ants collect flattened stalks from acacia trees and build a stockade around their nest entrance.*

particular direction, although this varies from colony to colony. If the bees are moved elsewhere, they will orientate their new combs in the same way as the old ones.

One social insect that refuses to be put off by changes in its environment is the mulga ant, another inhabitant of Australia. Its nest site has a distinctive circular stockade surrounding the entrance to its system of underground chambers. This fence, about 6 in (15 cm) wide and a similar height, is made up of thousands of flattened leaf-like stalks from acacia trees, known as phyllodes. The function of the fence has been somewhat of a mystery. However, it is assumed that the wider-than average, funnel-shaped entrance to the nest (about 20 times the width of the ant) would

be prone to flooding, and the fence is the mulga ant's flood-protection scheme.

One of the most remarkable builders is a tiny creature quite new to science. The bauble spider was discovered in 1994 in the Cedarberg Mountains in South Africa. Unlike its web-spinning relatives, the bauble spider constructs a 1/2 in (13 mm) diameter bauble of silk, which it suspends from overhanging rocks by a single thread. Inside the ball is a spiral tunnel, into which the 1/8 in (3 mm) long spider retreats. Some baubles are inhabited by a male and female, each with its own tunnel. Their tunnels, 1 2/5 in (35 mm) long, intertwine inside the bauble, while a central chamber contains the eggs.

FIT AND ACTIVE 3

JUMBO BRUSH *Elephants use tools such as creepers and bushes to remove flies and parasites.*

CHIMPANZEES AND BEARS ARE KNOWN TO TAKE MEDICINES FROM PLANTS, RED DEER OBTAIN CALCIUM FROM THEIR NEIGHBOURS, WHILE ELEPHANTS, GORILLAS, MOOSE AND MACAWS SUPPLEMENT THEIR DIETS WITH ADDITIONAL HEALTH-PROMOTING MINERALS FROM SALT-LICKS, CLAY-LICKS AND FROM THE SOIL. GREBES AND OTHER FISH-EATING BIRDS EAT ROUGHAGE WITH THEIR FOOD AND TAKE ADDITIONAL MATERIAL BETWEEN MEALS. IN THE MAIN, MOST ANIMALS HAVE HEALTHY, WELL-BALANCED DIETS, AND THEIR BODIES ARE KEPT IN PRIME CONDITION AND HIGHLY TUNED IN ORDER TO REMAIN FIT AND ACTIVE. IF THEY DO NOT, THEN ALERT PREDATORS ARE WAITING, PROGRAMMED TO SPOT THE UNFIT, THE INFIRM AND THE INJURED.

FRESHENING UP *A white egret preens its feathers.*

HEALTHY BODIES

Dietary supplements, sugar-rich diets, sanitary nest-wear, toothcombs, showers, exfoliation, toning, moisturising and mud-packs – animals make use of a cornucopia of health products in order to keep their bodies in good condition.

An unhealthy animal is a dead animal. In order to survive the rigours of a hostile world, in which an individual must fight for territory and the right to mate, be quick to take advantage of every feeding opportunity, and must be able to escape from predators, animals need healthy bodies. They therefore take body maintenance very seriously and assign many valuable waking hours to ensuring that everything is in good working order.

Changing your coat, or even changing your skin, is a highly effective way of keeping clean and parasite-free. Snakes and lizards slough their skins regularly, sea snakes going through the most extraordinary contortions to scrape off every flake and every last barnacle. Belugas or white whales moult their tired and yellowing skin each year, in the same freshwater estuaries in which they drop their pups. Inuit hunters tell of white whales with yellowing skin entering estuaries in the spring and leaving again pure white. The whales swim onto gravel banks on the rising tide and rub the old skin off on the stones. This can be dangerous, for if they miscalculate and roll in the shallows on a receding tide, they can become stranded and vulnerable to attacks by polar bears. Killer whales frequently visit traditional 'rubbing' beaches to scrape their bodies on the pebbles.

COLD COMFORT BEACH
A killer whale rubs its body on a pebble beach to remove dead skin and external parasites.

Among land-based mammals, some, such as monkeys and apes, groom each other, while herbivores use their teeth to groom themselves. It is always tempting to assume that the size and shape of an animal's teeth are related to the way in which it feeds, but this is not always the case – as the impala, a small African savannah antelope, demonstrates. The impala has a narrow muzzle with four centrally placed, spade-shaped incisor teeth – two at the top and two at the bottom – at the front of its mouth, which it uses for browsing on young leaves and shoots. On either side of these is another incisor and then the canine teeth, all of which are narrow and pointed and not well embedded in the jaw. These are thought to be toothcombs, positioned at the front corners of the mouth so that when the animal bends its head back, it is able to groom its body, removing water, dirt, ticks and other parasites, and possibly spreading the animal's own secretions, such as sexually attractive scents.

During a bird's life, individual feathers are damaged or lost, and are replaced immediately. But each year birds also moult all their feathers, which are renewed in a process that varies from species to species. European garden birds moult several times a year, shedding key flight feathers from the right wing and then the left, in such a way that they are not lopsided: an imbalance of feathers would prevent a bird from taking off and leave it vulnerable to predators. By losing feathers gradually, temperate garden birds ensure that they can forage, escape danger and keep warm, albeit at a reduced level of efficiency. Large birds of prey may take a couple of years to renew their

plumage, moulting feathers continually in order to stay airborne and able to hunt.

Some birds take advantage of circumstances to moult their feathers all at once. The female hornbill, for example, is incarcerated in her nest hole for the duration of the breeding season, so she moults all her feathers at once while safely out of the way. Water birds also tend to moult all their feathers in a short period, which makes them vulnerable to predation. To counter this, pink-footed geese in Iceland gather together

BALANCED DIET *Great crested grebes supplement their diet of fish by eating their moulted feathers for roughage.*

FEATHER CARE *A white egret from North America keeps its feathers in good condition by frequent preening.*

in large flightless groups and seek the safety of lakes; and over 100 000 shelduck, representing almost the entire population of north-west Europe, congregate each year in the Grosser Knetchsand area of the Waddenzee in the Netherlands to moult.

Great crested grebes eat their moulted feather debris as roughage, the number of feathers consumed depending to a large extent on the type of fish that the birds have been catching. Grebes feeding on bony and spiny fish, such as pike and perch, which contain a lot of indigestible bones, need to eat fewer feathers than those feeding on more easily digestible fish, such as smelt, because feathers and fish bones serve the same purpose: they form the nuclei of pellets that can be regurgitated, thereby, it is thought, helping to keep down the number of gut parasites, such as nematode worms. It is, perhaps, significant that grebes are unique in continuing to moult belly and flank feathers during the winter.

TOXIC SKIN

Parasites are the scourge of most animal species, and many animals have found ways in which to keep their live-in and live-on livestock to a minimum. It has been known for centuries that toads and salamanders have poisonous glands – toads with their warts and the salamander with its glandular ridge – and frogs such as the South American arrow-poisons have skins that exude poison strong enough to kill people. It was thought at one time that these poisons – the strongest known poisons in the natural world – were a part of the amphibian's

POISONOUS SKIN *The skin of the yellow-bellied toad secretes antibiotics that keep it free of bacteria.*

OCHRE TINTING *South African lammergeiers change colour after a dust bath in iron-oxide-rich potholes.*

defence against predators, but new research has shown that these chemicals have antibiotic properties and this is now thought to be their primary purpose.

The poisonous yellow-bellied toad, for example, has antibiotics in its skin secretions strong enough to kill a wide range of bacteria, including waterborne bacteria that cause gastroenteritis in humans and reptiles. The toad inhabits stagnant waters and is therefore prone to skin infections, which it fights with the secretions from its skin. Similar antibiotics have been found in the poisons produced by three widely differing creatures: the European honeybee, the cecropia moth and the African clawed toad.

Some birds also produce poisons. Three species of the strong-smelling thrush-like pitohui from New Guinea have toxins in their feathers. The discovery was made quite by accident, when ornithologists studying the hooded, rusty and variable pitohuis handled the birds and then put their fingers to their mouths. They felt a numbing and tingling sensation in their

lips. Curious to know what had caused the numbing, they analysed the tissues and found that the culprit was none other than homobatrachotoxin, one of the poisons also secreted by the arrow-poison frogs of South America. Although the strength of the chemical in the feathers is about 1000 times weaker than in the skin of the frogs, it is, nevertheless, effective against feather lice in adult birds and deters reptilian predators, such as snakes, which would otherwise develop a fondness for pitohui chicks.

The lammergeier or bearded vulture does not have poisons in its skin, but it too has an unusual way of taking care of skin and feathers. The bird normally has slate-grey wings and a purplish hue to its body, but head, neck and underparts are often a bright orange-red colour: that is, if it is not raining, when the coloration becomes paler. The colour is not permanent; the bird simply acquires a dusting of iron oxide from the

BIRD BATHTIME *A house sparrow splashes about in water to keep its plumage clean.*

mountain potholes and ledges where it spends much of its time. The pigment is spread evenly through the plumage when the lammergeier preens, and serves two purposes: it repels feather lice, and provides nesting birds with some degree of camouflage at their cliff-top nest sites.

Close examination of the lammergeier's outer body feathers shows that the tips of worn feathers are coated with a blob of iron oxide. Each feather has, in effect, a protective cap, which reduces the risk of further damage – the human equivalent of a remedy for split ends.

BODY CARE

Feather care is important to a bird, just as servicing a passenger jet is important for it to stay in the air, because if birds do not maintain their feathers they will fall out of the sky. A daily bath is one way to look after plumage: sparrows splash around on the edge of a pond; babblers jump in and out time and time again; and frigate birds take a quick dip when skimming the water surface. Kingfishers plunge in; hummingbirds fly through waterfalls; parrots like rain-bathing; and hornbills make use of wet vegetation and 'foliage-bathe'.

Preening is another daily toilet routine, the oil coming from glands under the tail.

It was thought at one time that oiling simply gave the feathers a degree of water-proofing, but it is now known that the oil also has antibacterial properties and acts as a fungicide. Anting, in which a bird grasps

---

SHARKS DON'T GET CANCER

Sharks are known to be resistant to cancers. Indeed, during the 20 years that the Smithsonian Institute in Washington D.C. has kept records of shark tumours worldwide, only six have been recorded among the hundreds of sharks examined. The secret seems to lie in the large amount of cartilage in the shark's skeleton. Cartilage contains a key chemical, known as an 'angiogenesis inhibitor', that prevents the growth of new blood cells. Cancers grow by forming their own nutrient-supplying blood vessels around themselves; the inhibiting factor in shark cartilage stops the growth of these blood vessels and thus restricts the food supply to the malignant cells.

---

a formic-acid-squirting ant and runs it over the plumage, is a rather drastic way to get rid of parasites, as is smoke-bathing: rooks, crows and other corvids can be seen with their wings outstretched over smoking

chimneys; the smoke, it is thought, causes external parasites, such as feather lice, to let go and drop off. Dust-bathing is another useful way of keeping clean and parasite-free, as dust stifles parasites and absorbs contaminants and excess preening oil. A sparrow living near a British factory canteen even took to sugar-bathing, in the absence of suitable dust.

However, birds are not the only creatures to use dust-bathing as part of their cleaning routine. Elephants, along with rhinos and warthogs, enjoy rolling in mud; and a liberal coating of dust after a mud bath will stifle lice. Elephants have also earned a place in that exclusive club of animals which use tools. Unlike chimpanzees, which use tools for collecting food, elephants focus more on personal hygiene. They collect branches and leaves for getting at all those private little places, for swatting flies and generally for scratching irritating parasite bites.

Animals are at their most vulnerable to parasites and disease in the home, with its

BEAUTY TREATMENT *A white rhinoceros enjoys a cooling mud bath. The mud smothers parasites and prevents others from gaining a foothold.*

closely packed bodies, food scraps and dead skin. And no home is more susceptible to infestation than a bird's nest, because all the activities are confined to a relatively small, cup-shaped space. Many birds solve this problem by lining their nests with medicinal plants that act as a natural, ozone-friendly insect-repellent. Starlings, for example, line their nests with madder, chrysanthemum, asters and other plants, which produce volatile compounds that kill or deter parasitic insects.

A song-bird parent, such as a bluetit, spends much of her day flying to and from the nest, foraging for caterpillars and feeding her young; but on her last flight of the day, she brings back a bunch of ground ivy, wood sage or stinging nettles with which to fumigate her nest before settling down for the night.

DIETARY SUPPLEMENTS

In places where key compounds are missing from the soil, and therefore from food plants, animals need to take dietary supplements. The plant-eaters encounter the most

ANTS WITH A HEAD START

The leaf-cutting caste of leaf-cutter ants are plagued by parasitic flies. A female fly injects an egg into the ant's head, and the fly larvae eat away at the large jaw muscles until the ant's head is an empty shell. The flies, however, are only active by day, and only land on heads that are more than 1/16 in (1.6 mm) in diameter. The ants, therefore, have changed their behaviour in order to thwart the flies. Ants with heads smaller than this forage by day, whereas those with bigger heads are more likely to appear only at night.

problems with deficiencies in their diets, but even carnivores need supplements. When a dog or cat eats grass, it is probably obtaining folic acid, one of the B vitamins necessary for manufacturing body-building proteins.

Red deer on the Scottish island of Rhum, and Shetland sheep on Foula, not

KILLER DEER *Red deer on Rhum rectify a calcium deficiency in the soil by killing Manx shearwater chicks and eating their bones.*

only take the occasional insect or snail with their normal daily diet of grasses and herbs, but they also catch and eat larger animals, the deer biting off the heads, legs and wings of Manx shearwater chicks, and the sheep decapitating the chicks of Arctic terns. As a result, the red deer on Rhum are the island's main predators, while eagles, ravens and crows have been reduced to scavenging bystanders.

The deer and sheep are very selective in what they eat. Unlike more usual carnivores, which eat flesh and entrails, they are only interested in those parts of the chicks' bodies that have high concentrations of bone. The clue to this strange behaviour lies in the fact that the island vegetation is low in calcium, and deer particularly need calcium when approaching the time of the rut, when the bodies of both hinds and stags must be in peak condition. Elsewhere, stags have been seen eating bones, discarded antlers and mineral-rich vegetation, but on these remote islands there are only seabirds to provide the dietary supplements. The shearwater chicks emerge from their burrows to exercise and learn to fly from late August to late September, precisely the time of year when red deer are preparing for the rut.

Most herbivores have not taken to killing, but visit salt or clay-licks for mineral supplements. At Kitum Caves, about 7800 ft (2400 m) up Mount Elgon, in East Africa, herds of elephants have excavated an entire cave system in their quest for the minerals sodium, calcium and magnesium. As a result, the caves, which also contain waterfalls, roosting fruit bats and nesting birds, form an enormous salt-lick. It is thought that generation after generation of elephants have excavated the caves from

scratch, a process that has probably taken about 100 000 years.

The elephants need to visit the caves because heavy rainfall in the area leaches minerals vital to the elephants out of the soil, and these minerals are therefore absent from the plants on which the elephants feed. In the caves, the elephants gouge out chunks of sodium sulphate from the walls using their short, blunted tusks, pick up the chunks with their trunks and crush them between their great molar teeth.

It is not known how often individual elephants go there, but the caves are visited by elephants almost every night. The mines are so deep that the visitors have to find their way in the dark, the skeletons of baby elephants providing proof of the danger of falling into crevasses. On average, elephants remain for four or five hours, eating salts for half an hour and then sleeping. In the mountains the nights are cold, but the temperature in the caves remains a comfortable 14°C (57°F), and so the dozing elephants are joined by bushbuck, waterbuck, duiker, buffalo and baboons, all coming to collect their ration of essential minerals.

Like the Kitum elephants, the rare mountain gorillas of the Virunga Mountains on the Zaire-Rwanda border also visit favourite mineral sites, where they dig out and eat handfuls of earth rich in calcium and potassium, creating miniature earth caves in the forest. However, the most colourful assembly of animals at a mineral-rich site takes place in the tropical forests of Peru, where hundreds of noisy, vividly coloured macaws flock to exposed cliffs to obtain kaolin present in the clay there.

Such congregations are unusual, in that macaws tend to live in pairs or small family parties of three or four birds in the forest canopy, but the explanation seems to be

MINING FOR MINERALS
Elephants visit Kitum Caves in Kenya, where they dig out chunks of sodium-rich rock to supplement their diet.

that there is safety in numbers. They are very conspicuous at the clay-lick, and predators are quick to take advantage of the situation. Black-and-white hawk eagles fly low, hidden by the foliage of the mid-storey and under-storey, and ambush the macaws as they descend to the clay-lick. Many pairs of eyes have a greater chance of spotting them approach, and so one species of parrot, the large red-and-green macaw, will not leave the cover of the trees until at least 40 other red-and-greens are present.

The macaws have to balance the time they spend eating kaolin at the clay-lick against the risk of being caught, and individuals do this in different ways. Some find a prime position on the lick site, stay for six

or seven minutes and then fly back to the trees, which maximises their time at the lick but exposes them to the greatest danger. Others stay for a couple of minutes at a time during a half-hour period, thus reducing the risk but also reducing the amount of time spent licking clay.

Researchers are still unsure why macaws visit their clay-licks when it exposes them to such danger. However, the clay itself is rich in minerals, such as sodium and calcium; it is alkaline, which might counteract acid indigestion; and the kaolin present might help the bird to neutralise the toxins in seeds, thus enabling the macaw to exploit a food source denied to other, less resourceful seed-eaters.

OXYGEN, WATER AND TEMPERATURE

Some animals survive in the most inhospitable places on Earth – in soaring desert temperatures, in frozen wastes, or deep in the sea – each having adapted in its own way to the ecological niche that it has striven to occupy.

Animals need water, oxygen and a regular temperature in order to thrive, yet some creatures spend long periods of time, and even their whole lives, in the most extraordinary places, in environments where it is difficult, or impossible, to obtain oxygen, where there is little or no water or food, and where there is extreme heat or cold. They live in, or visit, these remarkably hostile places because they have little competition there, and they have, therefore, evolved the wherewithal to remain alive: adaptations which ensure that they can obtain sufficient air, water and food, get about, and cope with the extremes of temperature.

Breathing is fundamental to all living things. In the main, surface-dwelling animals have few problems in breathing, apart from the camel in a sandstorm, and even this peculiar creature has a trick up its nose. Camels are able to close their noses in order to keep out flying sand and dust. They are also able to keep water loss during breathing to a minimum by moistening inhaled air and cooling exhaled air in the nostril cavities, which contributes to their ability to go for a long time without water.

DESERT NOSE *The bactrian camel conserves water, even when it breathes. The secret is in the nose.*

Similarly, the saiga, one of the goat antelopes that lives on the cold, arid Asian steppe, has a swollen nose in which inhaled air is warmed and moistened, and downward-turning nostrils to keep out dust.

Burrowing animals or those living in tunnels, however, have a basic problem, since heat and exhaled carbon dioxide could build up to such an extent in those conditions that it could wipe out an entire colony. Nature, however, has found a solution: for example, termite mounds and prairie-dog burrows share a self-regulating air-conditioning system.

The prairie dog is a ground squirrel that inhabits the wide-open plains of North America. It lives in a complicated system of burrows with many entrances, known as a prairie-dog 'coterie', and has a simple but effective way of keeping the air inside the tunnels fresh. The prairie dog enters and leaves its coterie by two types of entrance: around half of the circular entrances are at the tops of mounds that are about 6 in (15 cm) high, whereas the others are about 18 in (46 cm) high with firm sides and a chimney-like top that resembles a miniature volcano.

As long as the chimneys are higher than the low mounds, air will be sucked gently through the tunnels at about 20 in (51 cm) a second, renewing the air in the coteries every ten minutes; since air outside the base of the chimney moves more slowly than air around the top, the resulting reduction in pressure pulls air through the tunnels. The chimney also serves as a useful lookout station.

A strain of Nigerian termites of the species *Macrotermes bellicosus* builds extraordinarily high towers, sometimes over 25 ft (7.5 m) high and up to 15 ft (4.6 m) across at the base. The tower is hollow and honeycombed with tiny pores, with the nest itself situated below. There are fungus chambers, in which the termites grow a special fungus on which they feed, nursery chambers for the larvae, wood stores, and the royal

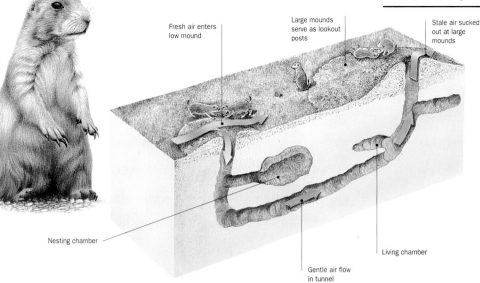

Fresh air enters
low mound

Large mounds
serve as lookout
posts

Stale air sucked
out at large
mounds

Nesting chamber

Living chamber

Gentle air flow
in tunnel

THROUGH DRAUGHT *The prairie-
dog burrow is built so that fresh
air enters at one of the tunnels
and stale air exits at the other.*

dioxide diffuses out. Oxygen-rich cooler air
from the basement is pulled up into the living chamber to fill the vacuum, and the cir-
culation continues automatically.

DEEP DIVERS

Air, or the lack of it, could also be a major
problem for those species, such as seals, sea
lions and penguins, that evolved on land
but have returned to the sea in order to
hunt for food. Although none of these creatures can breathe under water, they have
overcome problems caused by the effects of
water pressure and lack of oxygen, as well

chamber in which the queen, her consort
and guards reside.

The fungus gardens must be main-
tained at a constant 30.1° C (86.2° F) in
order that the fungus grows properly. However, there are so many termites going
about their business that all the
activity inside the nest could
cause the temperature to soar
uncontrollably above that; in
addition, the air could contain
as much as 15 per cent carbon
dioxide – enough to cause a
person to lose consciousness.
The secret of the termites' control over the situation lies in
the basement.

Below the nest, the termites
excavate an enormous cellar,
about 10 ft (3 m) in diameter
and 3 ft (1 m) deep. On the
roof of this chamber is a circular plate, supported by a column, on which the entire nest
sits, and hanging from the
plate are a series of narrow,
wafer-thin, concentric mud
vanes. The walls of the cellar
are covered in white crystals of
salts, indicating that evaporation has been taking place in
the chamber. Water permeates
down from the nest above into

the mud vanes and evaporates, cooling the
air in the basement, and oxygen filters in
through the thin nest walls. In the living
chamber above the basement, the warm air,
rich in carbon dioxide and low in oxygen,
rises into the high tower, where the carbon

AUTOMATIC AIR CONDITIONING
*Termite mounds, like modern
office blocks, have a system
that regulates the temperature
and recycles the air inside.*

EFFICIENT DIVERS *Seals and sea lions are able to dive to great depths and return to the surface without getting 'the bends'.*

– that dive to great depths and return rapidly to the surface with no ill-effects. How they do this was a mystery until a group of scientists attached a special tracking device to an Antarctic Weddell seal; this not only gave information about the depth at which the seal dived, but also sampled and analysed the composition of its blood on the way down and on the way back up. Their findings were quite a revelation.

The seal breathes out before it dives, and when it reaches a depth of about 100 ft (30 m) its lungs collapse. Any residual nitrogen – the gas responsible for the bends – is pushed into the upper respiratory tract and is therefore unable to pass into the blood; all the nitrogen already in solution is taken up by the body tissues, such as muscles, where it stays until the seal returns to the surface. With no nitrogen in the blood, bubbles cannot form as the seal ascends. At the surface the lungs expand, and the nitrogen in the tissues returns to the blood. Using this technique, the Weddell seal can dive very deep, with recorded dives of 2000 ft (610 m) lasting up to an hour.

Among the birds, the loon or great northern diver has been observed pursuing prey to depths of 220 ft (67 m) in Lake Superior, and a guillemot was recorded 240 ft (73 m) below the cod-rich waters off Newfoundland. But the bird that dives deepest in its search for food is undoubtedly the emperor penguin, and it has developed special modifications to enable it to do this. It has a streamlined body, coated with a thick layer of fat to minimise heat loss, and short, fine feathers; feet set well back that,

together with the tail, act as a rudder; and wings modified as flippers that enable the bird, literally, to fly through the sea. It can stay below for up to 18 minutes and reach a depth of at least 1300 ft (400 m). The closely related king penguin has been tracked down to 787 ft (240 m).

Emperor penguins have to make long journeys across the ice from their breeding grounds even to find waters where they can hunt. In a tracking experiment with emperor penguins in 1992, satellite tags were glued to the dorsal feathers of two sets of birds: those that had been relieved of incubation duties and were returning to sea during the harsh southern winter, and those that were foraging for food to feed their chicks during the summer. The Argos satellite was able to get a fix on nine birds 28 times a day, giving a fairly detailed picture of their movements, and confirming their remarkable adaptations to life both on the ice and in water.

During the winter months, the birds walked or 'tobogganed' on their stomachs over fast ice for 185 miles (296 km) or more, at speeds of $1/3$ mph (0.5 km/h), in search of polynias – ice-free stretches of water into which they could dive. They did not go direct, but stopped off at ice-cracks and seal breathing-holes, where they could dive in and hunt. Once at the polynias, they dived mostly in midwaters but sometimes close to the seabed at depths of 1300 ft (400 m), ate their fill – about 7 oz (200 g) of food – and returned to their mate after a journey that had taken about a fortnight in all. The other group of birds, travelling over light pack ice and not needing to get back to an incubating mate, made even longer journeys – a round trip of about 1033 miles (1654 km) – in order to feed at the edge of the pack ice.

Though seals and penguins are deeper-than-average divers, the world record for deep diving must be held by the sperm whale. It is the deepest-diving mammal, and it has been established for certain, using underwater sonar, that individuals go down to at least 3937 ft (1200 m) – quite a feat considering that they can only breathe at the surface so have to suspend breathing

as the loss of body heat, during long, deep dives.

After a deep dive, a human diver returning rapidly to the surface, without decompression stops on the way, will almost certainly suffer from 'the bends'. The decrease in water pressure as the diver rises causes nitrogen bubbles to form in the blood; this is accompanied by considerable pain, particularly in the joints. If the diver does not undergo an immediate programme of decompression, either by returning back down and rising slowly, or by entering a decompression chamber, then he will die.

However, there are many marine mammals – seals, sea lions, dolphins and whales during a long dive. And there is circumstantial evidence that they go even deeper. A bull sperm whale, thought to have been feeding on the bottom 10 500 ft (3200 m) below the surface, was caught and found to have two bottom-dwelling sharks in its stomach. Several whales have been found entangled in submarine cables hauled up from 3740 ft (1140 m).

Why the sperm whale should want to dive so deep has never been established, although it is suggested that it is utilising a food resource denied to its competitors. Seals, dolphins and the larger fishes skim off the squid found in surface and mid-waters, while the sperm whale hunts the inky depths for large deep-sea squid, including the giant of all giants, the giant squid, *Architeuthis*.

Such a large creature might be expected to have problems moving, and especially diving, in water, but as far as researchers can ascertain the sperm whale dives almost straight down at speeds up to 558 ft (170 m) per minute. Bulls dive the deepest and the longest, sometimes for up to two hours. They return to the surface almost as fast as they go down and avoid the bends by using the same method as Weddell seals. The absorption of oxygen and nitrogen in the muscles, however, is even more efficient.

A whale might dive many times with only a few minutes' pause at the surface, although there comes a time when it must rest. It is able to move rapidly through the water using its powerful tail, but its rapid descent and ascent are thought to be the result of an unusual anatomical feature. The proposal is only a theory, with little evidence to back it up, but it is our only explanation of the diving abilities of this remarkable creature.

The sperm whale's characteristic shape is the result of a large mass of tissue, known as the spermaceti organ, in the front of its boiler-shaped head. It is thought that one function of the organ is to act as a sound lens, focusing very high frequency sounds into a sound beam in front of the whale's head. The sounds bounce off objects in the water and return to the whale, giving it information about the distance, direction of travel, and composition of, say, prey

DEEP FLYERS *Emperor penguins can 'fly' down to depths of 1300 ft (400 m) and return rapidly to the edge of the pack ice, where they seem to shoot out of the sea.*

animals. It has even been suggested that the sounds emitted can be of such an intensity they can immobilise prey, and even kill it. There is, however, another proposed function, which relates more to diving than to hunting.

The spermaceti organ is filled with a special wax that has the property of melting at 29° C (84.2° F) precisely, and it is permeated with a complex plumbing system of blood vessels, sinuses and nasal passages. The proposal is that by causing the wax to melt or solidify in a controlled manner, the whale is able to regulate its density in the water and influence its ability to go down or up. At the surface, seawater is circulated to cool the wax, which shrinks and becomes more dense, causing the whale to sink. At the end of the dive, body heat generated by the muscles is carried

SNOWMOBILES *Tobogganing across the ice, emperor penguins go in search of areas of ice-free water in which to dive and hunt for fish.*

by the blood into the spermaceti organ where it melts the wax, causing the head to be less dense than the water, and allowing the whale to float to the surface with the minimum of effort. No matter how exhausted the whale becomes at the bottom, chasing squid or deep-sea sharks and zapping them with sound, it is assured a safe journey back to the surface.

POND LIVING

At the other end of the size scale are the aquatic insects and other invertebrates. They, too, must breathe air at the surface, but because of their small size they have a fundamental problem. They must break through the film on the water surface, which, for an animal the size of a mosquito larva, for example, is like a tough, almost impenetrable membrane. The miniature aquanauts have solved the problem in a variety of ways; many have developed devices designed to pierce the surface film, while others carry pockets of air below.

Most species of mosquito larvae have a 'snorkel' which they push up through the surface film into the air, although there is one species that rasps away the tissue of plant stems with its tooth-covered, cone-shaped snorkel and sucks out the oxygen from inside the plant. The water scorpion has a long breathing tube, or tail, on the end of its abdomen, resembling the sting of a land scorpion, that enables it to stay down and watch for prey. It grasps a water weed and pushes the tube upwards into the air above in order to breathe without having to swim to the surface.

The $1/2$ in (1.3 cm) long aquatic larva of the drone fly, known as a rat-tailed maggot, has a telescopic siphon that channels oxygen into the gills at the maggot's rear end, and that can be extended by as much as $2^1/_2$ in (6 cm).

The water spider, a small brown spider looking much like any other spider, spends most of its life under water. It breathes with the help of its own diving bell, which it builds by secreting a silken blanket between water weeds and collecting air bubbles beneath it. The spider takes down the air from the surface, bubble by bubble, between its back legs and abdomen. As the air builds up under the silk, it bulges upwards in a convenient air-filled cavity. The spider is able to wait inside the bubble until its quarry passes by, when it darts out and catches it. As the air supply gets used up, the spider replenishes it by adding more bubbles.

WING SCALES

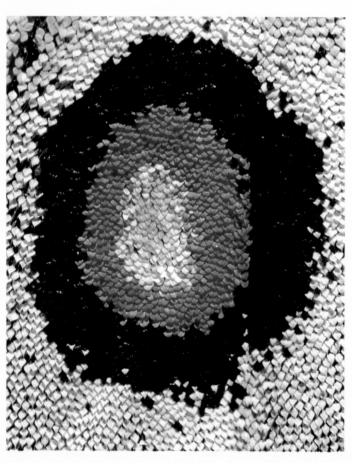

The tiny scales on the wings of moths and butterflies have several functions. They provide the insect with its own colour pattern and improve lift during flight. The loose scales on the feathery wings of the plume moth help prevent it from sticking to the leaves of the insect-catching sundew on which it lays its eggs, and the detachable scales on other moths and butterflies help them escape from spiders' webs into which they may have blundered. There is, however, another important function: the scales are also involved in insulation and heat regulation.

Different families of butterflies and moths have wing scales of different sizes and shapes. In general,

WING TILES *The overlapping wing scales on an apollo butterfly's wing produce identifying colours and patterns, and also help the insect escape from spiders' webs.*

butterflies tend to have a thin layer of small scales, whereas moths have thick layers of large and small scales, and their arrangement is linked directly to behaviour.

Nocturnal moths, which warm up by vibrating their wing muscles and make rapid wing beats when flying, have a thick layer of scales at the bases of their wings that insulates the thickest section of their wing veins, the place where they would be prone to most heat loss. Day-flying moths, which are able to gain energy from the sun by basking, have reduced layers of scales to prevent overheating. Some even have transparent wings with a complete absence of scales.

Butterflies, which only fly by day, rely totally on the warmth from the sun, and have larger wings in order to absorb more heat. The scales are thinly layered, for when butterflies are active, that is, when the sun is shining, they have no need for thick insulation.

The water boatman, which rows across the undersurface of a pond with two legs resembling oars, takes its own air supply with it. The creature has no gills, but retains a bubble of air between lines of hairs on its abdomen; the bubble is in direct contact with the pores or spiracles through which the insect breathes. The water boatman has hairs on its rear legs with which it detects and intercepts mosquito and other insect larvae that must visit the surface periodically to breathe.

The great diving water beetle also remains mobile by carrying small bubbles, like the aqualungs on a scuba-diver. It charges up its 'aqualungs' by pushing its rear end out of the water and lifting its wing cases slightly. Air is drawn into two large spiracles and is trapped under the wing cases. A bubble also forms around the hairs at the tip of the abdomen. This self-contained air supply provides the beetle with enough oxygen to go hunting, usually for tadpoles, for about an hour.

STAGNANT WATER

There are times, however, when oxygen is just hard to come by. For example, fish living in stagnant water, where much of the oxygen gets used up by a superabundance

DIVING BELL   The water spider creates a bubble of air below the pond's surface, from which it can pounce on passing prey.

CUSTOM-BUILT SNORKEL
The aquatic larva of the mosquito breathes through an air tube or snorkel in its 'tail'.

of bacteria and algae, have evolved a variety of breathing mechanisms in addition to their gills that enable them to obtain enough oxygen to survive until the next flood re-oxygenates the water.

The gigantic pirarucu of the Amazon, one of the largest freshwater fishes in the world, often finds itself in water that is low in oxygen, and has a swim bladder that functions like a lung. The fish comes to the surface and takes gulps of life-sustaining air

into the swim bladder and oxygen is absorbed by the body

One fish family – the Characidae, of tropical South America – change shape to accommodate low oxygen levels at certain times of the year. During the driest months, when the swamps dry up and oxygen levels in the remaining ponds begin to get low, there is a real danger that the fish will suffocate. To avoid this, it develops a fluid-filled swelling at the tip of the lower jaw. Water at the surface of a stagnant pond, being next to the air above, has a little more oxygen in it than the water lower down. When the fish swims close to the surface, the swelling channels the oxygen-rich surface film into the fish's mouth without taking in any of

the deoxygenated water below. During a particularly dry period in northern Venezuela in the late 1980s, oxygen levels in residual ponds were so low that there was high fish mortality. The Characid fishes, however, were not among the casualties. Their curious anatomical adaptation had saved them.

African catfish in the Congo have solved the same problem in a slightly different way. The upside-down catfish, as its name suggests, swims belly up. Even its countershading – dark above and light below – is reversed. As it swims along it feeds on plankton and detritus in the surface film, but it can just as well feed on the bottom and swim the right way up.

When the oxygen level in the water deteriorates, however, the catfish automatically adopts the upside-down posture, swimming along the underside of the water surface. Like the Characid fishes, it funnels the top, oxygen-rich layer of water into its

AMAZON AIR-BREATHERS *The gigantic pirarucu survives in stagnant water by gulping air at the surface and absorbing it through its swim bladder.*

WARM CHICK

Inside its shell, a chick is dependent on its parents to keep it at a comfortable temperature. Most birds sit on their eggs to incubate them, but the white pelican stands on them. The parent checks the temperature with a patch of bare skin on the belly or through its webbed feet, but just before hatching, when it places the eggs (usually two) back between its legs and stands over them, it cannot monitor the temperature. Instead, the chick inside calls out to indicate whether it is comfortable or not. One or two calls a minute means that everything is normal, but an increased calling rate alerts the parent that the chick is too warm or too cold.

mouth and across its gills, excluding the water below, but it does so using the three sets of stiff, rod-like feelers on the tip of its lower jaw.

HOT DESERTS

Water, or lack of it, is clearly a problem for desert creatures. In some deserts that are close to the sea, such as the Namib of south-western Africa, where rainfall is very

rare except along the coast, mists and coastal fogs can extend as far as 50 miles (80 km) inland for about 60 days of the year, and some animals have adapted their behaviour to take advantage of this fleeting bonanza. The basic principle is that any vertical object placed in the path of the fog will provoke precipitation of the moisture held in suspension, and several animals take advantage of this piece of physics. The appropriately named headstand beetle heads for the tops of sand dunes, where it tilts its abdomen to the sky and stands almost upright on its head in order to collect the fine mists on its body surface and guide the water droplets into its mouth. Near by, the button beetle excavates long furrows that act as fog traps. The nocturnal gecko licks condensation off its large eyes.

Hot deserts are a test for any creature, but some species have found a niche that no other self-respecting animal would care to explore, and thrive in the extreme conditions. A snail with the tongue-twisting name of *Sphincterocheila* lives in deserts where the ground temperature is way above the lethal limit and there is very little water to be had. The snail has a pure white shell that reflects 95 per cent of the sun's energy. It does not absorb heat from the ground because the shell touches at few points, and the snail retracts into its shell when not active anyway. The thickness of the shell prevents water loss and the animal can throw a thick mucous membrane across the opening. In this way it can survive three years' dehydration.

Frogs, such as the trilling frog and water-holding frog of Australia, avoid the problem of drought altogether: they go underground and aestivate. Aestivation is a behaviour similar to hibernation that occurs in places where it is too hot to be active, but where to survive more might be required on the animal's part than just digging a hole and hiding in it.

After a period of rain, during

WATER COLLECTOR *Tiny channels in the skin of the Texas horned lizard direct surface dew into the animal's mouth.*

which desert frogs and toads undergo a burst of accelerated reproductive activity – attracting partners, mating, depositing eggs, developing into tadpoles and metamorphosing into adults, all in the space of a few, frenetic weeks – they prepare for the coming drought. Many absorb huge amounts of water, which they store in the bladder. The rounded, swollen body of the spadefoot toad, for example, may be 80 per cent water, but before the last drop of water in the pond evaporates the frogs and toads

HOT HEADS

Some sharks have warm muscles despite being 'cold' blooded. The swimming muscles of the great white shark, mako and porbeagle are kept 7-10° C (12.6-18° F) warmer than the surrounding water. Tuna have warm muscles too, and can swim at speeds over 40 mph (64 km/h), particularly when pursued by swordfish. The swordfish swims in short, swift bursts in pursuit of fast-moving prey. It does not have warm muscles, but it has a heater in its head. In order to spot and chase passing food, particularly in the cold depths, it can heat up its eyes and brain to 14°C (57°F) warmer than the surrounding water, an ability it shares with white marlin and sailfish.

must find a place to hide. A frog will burrow backwards into the mud of its rapidly receding pond and dig down up to 3 ft (1 m) below the surface. Here it secretes a

watertight, mucous cocoon and will sit out the drought for months or even years. Using such a strategy, the trilling frog can even survive in the sand dunes of Australia's Simpson Desert.

The water-holding frog, *Cyclorana platy-cephala,* from Australia, is the supreme desert survivor. It burrows down for 12-15 in (30-38 cm), where it excavates a chamber about twice its body size. It then shuts its body down by slowing its heart and breathing rates to a minimum. After it has been dormant for about two weeks, its outer skin cells separate from the underlying layer until the entire animal is covered by a loose envelope, save for two tiny tubes to the nostrils. It can survive in this state for over two years.

HIBERNATION

At the other extreme, cold can be a killer. Hibernation, during which the body shuts down almost completely, is a way in which to avoid being caught out in the cold. During hibernation the heart rate drops considerably, the body temperature is reduced and the animal's metabolism generally slows down, and it must hide somewhere where it is safe from the worst of the cold.

DESERT FROG *When it rains in an Australian desert, the water-holding frog breaks out of its cocoon of dead skin and sets about breeding.*

Hornet queens spend the winter hiding under the thick bark of a fallen oak, ash or elm, literally in suspended animation. Each autumn they dig out a small cell, not much larger than themselves, using their feet and jaws. The air pocket protects the insect from all but the hardest frosts and prevents it from being squashed. Wolf spiders, beetles and fly larvae postpone hostilities for the winter, and may find a communal hibernation site inside the ribbed stem of a hogweed plant. Adult lesser stag beetles join their wood-boring larvae in tunnels in trees that have died standing. The sawdust from the larvae, which take several years to pupate and develop into adults, is prime insulation against the cold.

Parasitic flies and wasps depend for their survival through the winter on their host dying in a suitable place for them to hibernate in – one where they will be protected from the worst of the cold. The

TEMPERATURE AND SEX

The sex of many species of hatchling reptile including turtles, tortoises, crocodiles and alligators, is determined by the temperature of the eggs during incubation. It is not already determined when they are laid.

Studies of American alligators have shown that at an incubation temperature of less than 30°C (86°F) only female hatchlings emerge, while at nest temperatures over 34°C (93°F) all the offspring are males. Mother alligators are able to regulate the temperature at which their eggs will incubate by choosing particular nest sites. Those on dry levees produce mainly male babies, while those in wet, marshland areas contain mostly females. In a population of alligators the usual ratio is about 5:1 in favour of females. This ensures that breeding stocks are not dominated by a few older females.

Sex determination in the eggs of sea turtles is the reverse of that in crocodiles and alligators. At higher temperatures, around 32°C (89.6°F), all the offspring are females, while at lower temperatures, around 28°C (82.4°F), the hatchlings are male. At 30°C (86°F) equal numbers of males and females are produced.

Snapping turtles are even more curious. They produce more females at higher temperatures and more males at lower ones, but at even lower temperatures females predominate again.

IT'S A BOY! IT'S A GIRL!
The sex of these baby American crocodiles is determined by the temperature of the nest.

thick-headed fly, an internal parasite of bumblebees, makes sure it hibernates in just the right spot by influencing the behaviour of its host. Normal bumblebee workers die on the ground, but those infected with the fly larvae bury themselves first. Usually, only bumblebee queens that have recently mated bury themselves in autumn and live through the winter in hibernation. It is thought that the parasite switches off the production of a special hormone in its host that suppresses sexual development. The infected worker therefore behaves like a

SUSPENDED ANIMATION
Queen hornets find a safe crevice during winter and shut down their bodies to conserve energy.

young queen and unknowingly digs its own grave. As a result, the parasite is buried safely below the ground, out of reach of winter frosts. In spring it can return to the surface and find a new host.

Dormice do not head for nests each autumn as was once thought. Instead they curl up among the leaf litter, under patches of moss or in the hearts of large ferns on the forest floor and go to sleep in little bundles of fur. They are not vulnerable to foxes, stoats or weasels because their body temperature is very low and they have no smell to attract predators. The dormouse needs to keep its body temperature low, but not too cold, for every seven to ten days it must wake to rid itself of body wastes created by the burning up of reserves of stored fat.

Most hibernating animals have a waste-disposal problem, although it is minimised because the bodies of many are just ticking over. Some emerge briefly to defecate or urinate and return to hibernation again. The North American black bear, however, has a different way of dealing with the problem. At the end of October each year, it retreats from the world, seeking out a suitable den where it will sleep for three months or more. Unlike other hibernating animals, the black bear does not drop its body temperature

down to the ambient temperature, but maintains it within 1-2 degrees below the normal active body temperature. This could be because black bears in winter dens are sometimes killed and eaten by grizzlies, so they need to be ready either to defend themselves or to get away in the event of an attack, or because the female gives birth inside her winter den. It also seems to be a time when the bear turns stored fat into muscle, which would keep its body temperature up. It refrains from eating and drinking for the entire period and does not urinate.

In the spring, when the milder weather causes it to wake up, the black bear, like all animals that hibernate, must make the transition between hibernation mode and normal, active mode. It does this gradually: a bear just out of hibernation wanders about aimlessly at first, showing no interest in food or water, until its body has caught up with its mind.

COLD-WEATHER MOTHS

Some creatures that, by all the laws of nature, ought to shut down to ride out the winter months are found flying about even when the temperature is close to zero. Many moths and butterflies do pass the winter as dormant larvae or pupae, but there are some species that have reversed their annual life cycle, becoming dormant in summer and active in winter. The main reason for this shift is that when migrating birds arrive in early summer to gobble up

HIBERNATING BIRD

The only bird known to hibernate is the common poorwill. It was first found on December 29, 1946, when an apparently dead bird was discovered in a vertical rock crevice in the Colorado Desert of southern California. Its body was cold and still and there was no perceptible heartbeat, yet, as it was placed back in the crevice, the bird opened one eye. Although modern science was late in recognising the bird's overwintering strategy, it had not gone unnoticed among the local Hopi Indians, who call it 'the sleeper'.

as many insects as they can find, the winter moths have gone to sleep. They emerge again in the autumn, when the birds have gone.

Winter moths, such as the sallow moths of North America, emerge from their pupal cases in the autumn, not the spring, and are active throughout the winter. They must keep the flying muscles in the thorax at a temperature of at least 30°C (86°F) in order to fly and they do this, firstly, by shivering for about 25 minutes, to bring up the body temperature in order to take off; secondly, by restricting, with a barrier of air sacs, the amount of heat moving to the abdomen, which would otherwise be lost to the outside. In addition, they possess a heat-exchange system in the main blood vessel in the thorax that maintains the heat in the wing muscles; lastly, they have a thick layer of insulating hairs on the thorax.

Their behaviour during the winter also helps them remain active. Instead of perching in the open, for example on tree trunks, as other moths do, they hide under loose leaf litter, where the temperature can be only −2°C (28°F) even though the air temperature is below −30°C (−22°F). In the spring, however, they are at a disadvantage and are prone to overheating. When the

body temperature reaches 20°C (68°F) they must stop all activity, and so they aestivate until the autumn.

ARCTIC ACTIVITY

Some insects rely on an ability to resist freezing, or to survive being frozen, in order to live in very cold conditions. This ability allows the thumb-sized, orange-brown woolly bear caterpillar of a lymantriid moth, for example, to survive in the hostile and barren High Arctic. And it lives for 14 years, a long timespan for an insect, 13 years of which are occupied by the third to sixth larval stages and pupation into the adult moth.

In general, insects, and indeed other animals living in polar conditions such as Antarctica's ice fish, survive freezing temperatures by lowering the freezing point of their blood and body tissues with the help of an antifreeze agent, such as glycerol or ethylene glycol, produced by the body, in the same way that a motorist protects the water radiator of a car.

The alternative strategy is to be freeze-tolerant, which involves preventing ice from forming inside cells while ice still fills the rest of the body, including the blood, gut contents, and the spaces between cells. The same chemicals that behave as antifreeze

agents, such as glycerol, also serve freeze-tolerant animals. In the latter case the agent helps to reduce the proportion of water locked up in ice, and thus tissues remain undamaged. With such a system, the woolly bear caterpillar can tolerate temperatures as low as −70°C (−94°F) during the coldest, harshest winters.

In summer the caterpillar tracks the sun, its hairy body absorbing and retaining heat, although it must ensure a healthy balance between warming up in order to feed and being cool enough so that the amount of energy it loses through normal bodily functions, such as respiration, is not greater than the energy it gains from the food. This juggling with temperature is reflected in the fact that the caterpillar is more efficient at metabolising food when its body temperature is at 16°C (60°F) than at 29°C (85°F). When the temperature drops below 4°C (40°F), however, the caterpillar starts to produce glycerol, and is therefore prepared in advance for any potentially damaging drop in temperature, summer or winter.

ARCTIC INSECT *This adult Arctic moth lives for just a few summer days, but its larvae may remain as caterpillars for 13 years.*

GETTING ABOUT

Animals have developed an extraordinary diversity of ways in which to get about. There are walkers, runners, crawlers, wrigglers, slitherers, squirmers, fliers, burrowers, swimmers, divers and many more besides.

Plants manufacture their own foods and can therefore stay put, whereas animals have to move in order to feed. Herbivores must find clumps of vegetation, and often move with the seasons, taking advantage of periods of plenty. Predators must move to catch their food, and the prey must move in order to get away.

Mankind may have invented the wheel, but there are examples in nature of creatures that can roll along. The golden wheeling spider from the Namib Desert lives and hunts on the sand dunes. It feeds on small insects, such as silverfish and beetles, which it catches from its trapdoor lair in the sand, but if it is not careful this predator becomes prey itself. It is plagued by a female parasitic wasp, one of the 1¼ in (3.2 cm) long tarantula hawk wasps, who lays her eggs in the spider's body, and whose larvae eat the spider from the inside out. Movement over sand can be slow and so, to avoid the wasp's inconvenient egg-laying habits, the spider makes quick its escape by tucking in its eight legs and rolling, just like a wheel, down the face of the sand dune.

Another roller is a tiny, 1 in (2.5 cm) long marine crustacean, *Nannosquilla*. It also lives in the sand, not in desert or dune, but under the sea off the coast of Panama, and it has lost the ability to walk. In the shifting sand environment in which it builds a mucous-lined tunnel, a narrow burrow is not only more stable than a wide one, but it also uses less energy to construct. The evolutionary consequence is that the body of this creature has become very elongated, with three pairs of ineffective legs at the front end. If beached, the creature has one course of action left: it cannot crawl or wriggle, so it curls up its body and executes up to 40 backward rolls at a time.

Another example of a wheel in nature must have evolved hundreds of millions of years ago. The flagellum on certain species of bacteria and on animal sperms is driven by a reversible rotary motor. It works in the same way as an electric motor that is plugged into a constant-voltage outlet. The difference is that the electrical supply comes not from a conventional generator but from the nuclei of hydrogen atoms. With this miniature motor, linked to a flexible flagellum, the bacterium or sperm is able to propel itself through water. It was, perhaps, one of nature's best kept secrets.

WALKING, CRAWLING, SLITHERING

In general, the wheel is virtually useless as a means of locomotion in the wild. Uneven terrain, boggy ground, shifting sands, boulders and logs hamper a wheel-driven machine, and engineers have turned to the insect principle of walking on six legs, or the spider's eight legs, as a basic design for mechanical robots, such as the eight-legged Dante used by volcanologists to 'walk' into the mouth of the active Mount Erebus volcano in the Antarctic. Creatures with two and four legs are inherently unstable, for when the animal moves, at least one leg on one side is always off the ground, causing the animal to tip in that direction. Spiders and insects, with eight and six legs respectively, have support on both sides with every step; in effect, they are stable platforms.

The ancestors of insects were worm-like

SPIDER WHEEL *The golden wheeling spider has a defensive display that deters attackers. It also has a unique escape strategy – it curls up and rolls.*

creatures and most probably had more pairs of legs, but they gradually reduced in number until six legs, the minimum needed for stability over a wide range of speeds, became the successful evolutionary trend. The insect concentrated its legs, and also its wings, on the thorax, the part of the body that developed into the animal's locomotor centre, and movement relied on the legs forming tripods.

It is difficult to see with the naked eye, but high-speed photography has revealed that the insect moves its legs in a particular pattern. The insect supports its weight with the first and third legs on one side and the second leg on the opposite side, to form a tripod. The remaining three legs move forward together, and then they too form a tripod, the insect moving forward on alternate triangles of support. This is very stable, and the creature can stop at any point in the cycle without toppling over.

Amphibians and reptiles have four legs, and usually use all four to get about. The frilled lizard of the savannah woodlands of Northern Australia, however, although a quadruped, spends much of its time on two legs. It is well-known for its threat display, during which it raises its body onto its hind legs and expands a frill of skin around its neck to form a ruff. If the predator is unimpressed, the lizard makes off. At first it runs on all four legs, but changes to the bipedal mode with the head slanting forwards when it has got up to speed.

More usually, however, the lizard spends its life in trees, not on the ground, descending to move to another tree, interact with others of its kind, or catch the occasional ground-living insect. When foraging on the ground, the lizard moves around on its hind legs more often than was first thought. Moving at a slower speed, it stands more upright, the long neck and frill enabling it to keep its centre of gravity over its hind legs, and from this erect position it has a better view of the lie of the land and can spot prey more effectively.

A relative of the frilled lizard, the aptly named 'Jesus Christ' lizard, or basilisk, of Central America, not only runs on two legs at speeds up to 7 mph (10 km/h), but also

runs on water. Again, it uses its unique skill to escape predators, such as small spotted cats, which patrol the river banks favoured by the lizard. It cannot perform its 'miracle' for any great distance, but can run far enough across the surface of the river to be safe. After a few strides it sinks and swims towards the other bank. The basilisk is able to walk on water with the help of its large feet and powerful back legs. The long, widely spread toes are fringed with scales,

THREATENING COLLAR
The frill-necked lizard of
Australia frightens predators
by raising its neck frill and
hissing.

and the legs move with an action not unlike the paddles of a paddle steamer, enabling the lizard to be buoyed up briefly by the surface tension of the water.

Some insects also have ingenious escape

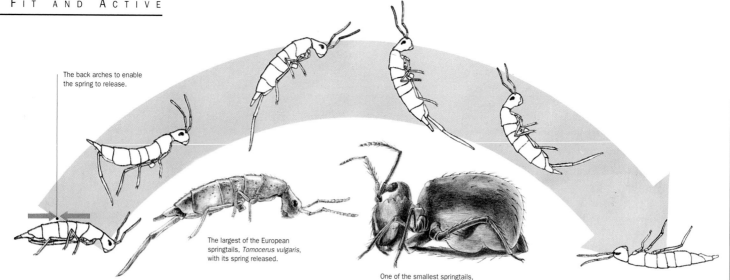

The back arches to enable the spring to release.

The largest of the European springtails, *Tomocerus vulgaris*, with its spring released.

One of the smallest springtails, *Dicyrtoma fusca*, in the resting position.

ESCAPE JUMP *The springtail, as its name suggests, has a spring in its tail which enables the insect to catapult out of the way of marauding wolf spiders.*

mechanisms, including ones that enable them to jump out of the way of prey. Springtails are extremely common insects – over 300 species occur in Britain alone – and they live in the leaf litter on the forest floor. They are no more than 1/4 in (6 mm) long, and usually get about by crawling on their six legs. However, if threatened by a predator, such as a wolf spider, they can leap high into the air, for at the rear end they have a spring-like tail, hence their common name.

The tail is normally folded under the body, like a closed safety pin, but when danger appears the springtail releases the spring, catapulting itself in an upward and forward trajectory at tremendous speed. At the same time it arches its back, like a human gymnast, and somersaults backwards, the result of having its spring at one end of the body. It lands in an undignified heap and scuttles away.

The 1/8 in (3 mm) flea can leap 13 in (33 cm) from one host to the next. It does this with a special spring, a modified wing hinge, which this now-flightless insect inherited from a flying ancestor. The spring

SIDEWAYS SLITHERER
The sidewinder rattlesnake of North America moves smoothly over sand by wriggling sideways on its ribs.

stores energy as the flea bends its body when preparing to jump, and releases its energy down into the legs, which propel the flea upwards. Similarly, the 1/2 in (1.3 cm) click beetle can project itself upwards as much as 12 in (30 cm) by jackknifing its body.

Coming down to earth, there are animals that creep and others that slither, and they do so using all manner of 'feet'. The microscopic amoeba has many flowing 'pseudo-feet'. How it moves these pseudopodia has been a matter for debate. Some scientists believe that the cytoplasm is squeezed forward, while others think that the cytoplasm at the front contracts and pulls the amoeba forward. There is even a theory that a ratchet system rotates the animal as it moves forward.

Snakes, legless lizards and certain salamanders progress by slithering. It was always

thought that animals using this form of locomotion use less energy than those that walk or run – that is, until scientists examined the black racer snake and looked more carefully at its movements. Moving close to the ground might seem less energy-dependent than being raised above it, but the slithering animal has other demands. A long body in contact with the ground encounters greater resistance than four feet. Muscles require energy to keep the ribs rigid and the head held up off the ground. All told, the slitherer uses just as much energy as the walker.

WATERSPEED

Animals immersed in water are more usually propelled by muscles and fins, or muscles and limbs, which, combined with a streamlined body shape, enables them to travel

CASUAL HITCHHIKERS
The bottle-nosed dolphin uses
very little energy when
it swims in the bow-wave
or wake of a ship.

easily through the water. But even here nature has been particularly clever; for example, a fast-swimming bony fish slips through the water because its body is the right hydrodynamic shape and is covered in a lubricating slime.

The shark goes one better: all over its body are tiny teeth, known as dermal denticles, that make it an even more efficient swimmer. This is because a smooth skin causes more drag due to friction, whereas the denticles break up the friction effect so there is less drag. Commercial aircraft manufacturers have even copied the principle, and aircraft with rivet-like protuberances all over the skin are proving to be more aerodynamically efficient than the traditional smooth-skinned types.

The dolphin is even more remarkable, for not only do the tissues beneath the skin

crumple and crease under changes in water pressure and flow, allowing it to slide noiselessly through the water, but also the animal sheds cells as it progresses: it quite literally slips out of its skin. This makes the dolphin ten times more efficient than humans at swimming. The dolphin, however, does not use up valuable energy unless absolutely necessary, and so will hitch a ride if the opportunity arises.

In tests in the USA, bottlenosed dolphins were trained to follow a boat while their heart and breathing rates were monitored. At the normal dolphin cruising speed of 7 ft (2 m) per second, about twice the top speed for a trained swimmer, the test animals exerted little effort in keeping up with the boat. Their heart and respiratory rates were no more than they would be at rest. When the boat speeded up, the dolphins worked harder only until they reached a speed of 13 ft (4 m) per second, at which point they stopped normal swimming, dropped back to the wake of the boat, and rode the trailing waves.

Wave-riding enables dolphins to travel at high speeds while expending little energy. Indeed, they hardly seem to twitch a muscle, only flexing their tail flukes to make minor adjustments in order to maintain their optimum swimming position on the wave. The practice was probably developed a long time before ships arrived, the resourceful dolphins using the

WATER FLYING   *The king*
penguin's wings are modified
into flippers, and the feet
and tail have turned into
rudders for tight turns,
enabling these large birds
to be highly manoeuvrable
underwater hunters.

wakes and bow-waves of large whales and the waves whipped up by the wind as a means of free travel.

Another energy-saving way to travel adopted by dolphins, in common with seals and penguins, is by leaving the water completely. The behaviour is known as 'porpoising', and is named after the dolphins' smaller coastal cousins. The animals periodically leap clear of the water, but they do not do it simply for fun. At a certain critical speed, depending on the size and shape of

HIGH-SPEED HUMMINGBIRDS

A hummingbird flaps its wings 80 times a second; to keep pace, its heart beats 1000 times a minute and it takes 250 breaths a minute. It can fly forwards at speeds of up to 47 mph (75 km/h), and uses more fuel per minute in relation to its size than a modern jet fighter. It utilises a high-energy fuel – the sugar in flower nectar – and may get 50-60 slurps per flower. It can beat its wings a million or more times without stopping. If a human body were subjected to such a high metabolic rate, it would heat up to a temperature of 399° C (750° F) and burst into flames.

the swimmer, the pressure wave formed in front of the head offers so much resistance that it would be counter-productive to try to swim harder. Instead, the animal uses the energy to push itself clear of the water. Once in the air it is faced with less resistance. By porpoising, penguins and dolphins

have the best of both worlds and can travel faster and further for the same amount of energy.

BELLS, BLADDERS AND JET PROPULSION

Fish have air bladders with which they are able to adjust their buoyancy. They can secrete oxygen from the blood into these bladders to fill them, like filling a balloon, in order to go up, or absorb oxygen from them in order to go down. However, it came as a bit of a surprise to find that a species of octopus has one, too. Octopuses are usually found at the bottom of the sea, among rocks and in crevices. They only swim to pounce on prey, usually crabs, or to escape from predators, by jet propulsion. But there is one species of octopus that has forsaken its rocky home and taken to the high seas.

This free-swimming octopus is *Ocythoe,* an animal seldom seen but which is known to be food for swordfish. It moves, like other octopuses and squids, by jet propulsion. It is able to contract its mantle and squirt water out of the siphon under its head. *Ocythoe,* however, has two additional nozzles, set high on each side of the body, which help it swim in a more controlled fashion, much as a spacecraft is guided by rocket jets in space.

The female of the species has a swim bladder, unlike other free-swimming cephalopods, such as the pearly nautilus

JET FIGHTER   *The day octopus of Hawaii can adjust its buoyancy to rise or fall in the open sea and moves quickly using jet propulsion.*

DEADLY BLADDER
The Portuguese man-of-war
remains afloat on the surface
of the sea with an air-filled
bladder.

and the argonaut, which have gas-filled, chambered shells. The lining of the swim bladder can secrete gases, much like the lining of the shell of the other cephalopods, enabling the creature to control its buoyancy and therefore its movement upwards and downwards. More gas in the swim bladder enables the octopus to rise in the water column, and less gas allows it to descend. Indeed, it is thought that the bladder derives from an ancestral shell, an organ that is still found in baby octopuses. The female requires the additional flotation because she retains her eggs in her body, making her heavier and more liable to sinking.

The Portuguese man-of-war also has a bladder to keep it afloat and drifts, not with the current but with the wind. The creature is not one animal, but many living together, with groups of similar individuals, known as polyps or 'persons', playing different roles.

Some specialise in digestion, others in reproduction, floating or catching food. The 4-12 in (10-30 cm) long float, which is frilled at the top, is filled with gas secreted by a special gland. Periodically the float is collapsed, first to one side and then the other, to moisten it and stop the membrane from drying out. When inflated it sits on the surface of the sea, where it is blown about the ocean by the wind. Under certain conditions hundreds of men-of-war are blown together in huge rafts.

The flattened, frilled part of the bladder is angled at 45°, and the vast array of stinging tentacles, which can reach down 30 ft (9 m) or more, acts as a sea anchor, helping to steer a course at a set angle to the wind. If a man-of-war in the Northern Hemisphere is caught in the westerlies, the sail aligns south-west to north-east, the tentacles trail to the north-west, and the creature makes progress to the south-east. This angling of the sail prevents the Portuguese man-of-war being blown into areas of doldrums such as the Sargasso Sea.

A relative of the man-of-war is the by-the-wind sailor, and it too sails in the breeze. Instead of a single, large, gas-filled bladder, the by-the-wind sailor has a circular disc of gas-filled chambers, 4 in (10 cm) in diameter, on top of which is a membranous 'sail'. It is colonial as well, with reproductive polyps surrounding a central mouth below the disc, and stinging polyps fringing the disc. The sail is also angled, so that the colony sails like a yacht, and flotillas of by-the-wind sailors travel together in the same direction.

Often accompanying them are purple sea snails. These voracious molluscs also float on the sea's surface supported by a raft of bubbles. They feed on the by-the-wind sailors, eating all the soft parts but leaving the circular disc, to which they sometimes attach their eggs.

Another unusual sailor is the southern right whale. It uses its tail, set like a spinnaker, and 'sails' with the wind just for fun. At Peninsular Valdez on the Patagonian coast the wind blows hard, and the fresher the breeze, the more excited these giant whales become. Sometimes one will stand on its head, lift its tail in the air and sail across the bay. Then it swims back to its starting place and sets sail all over again.

OUT OF THE SEA

Squid must keep swimming to avoid sinking to the bottom. They do this either by rippling their fins or by using jet propulsion, but a squid in a hurry may take to the air. Using jet propulsion from the directable siphon underneath its body, the flying squid is able to launch itself out of the water at heights of 10-20 ft (3-6 m) and cover a distance of over 50 ft (15 m). The flying squid is a frequent flier, and in doing so it joins an exclusive club of unusual flying animals, including fish, frogs, lizards, snakes, and squirrels.

Surging up to the surface at 15-20 mph

FAST FLIERS

The bird that reaches the highest speeds is undoubtedly the peregrine falcon. This remarkable bird of prey swoops on its victim, folding its wings in a near vertical dive and reaching speeds in excess of 112 mph (180 km/h). The fastest bird in powered flight, however, is considered to be the Asian white-throated spinetail swift, which has been observed to fly at 105 mph (170 km/h) during a courtship display. On more regular flights, ducks and geese are known fast fliers, often reaching speeds of 60 mph (97 km/h). Of the insects, hawk moths, horseflies and dragonflies are the fastest fliers. In tests, a hawk moth was timed at 33 mph (53 km/h).

(24-32 km/h), fins folded flat against the sides of the body, the flying fish prepares for takeoff. It leaves the water at an angle of about 15 degrees, spreads out its elongated pectoral fins and sculls momentarily across the surface of the water by vibrating the tip of its tail from side to side at a rate of 50 beats per second. It glides at speeds up to 40 mph (64 km/h) at a height of 4-5 ft (1.2-1.5 m), but after 10 seconds and a distance of up to 900 ft (274.5 m), airspeed drops to 20 mph (32 km/h) and the fish returns to the water. The flight need not end there, though, for a further flick of the tail will have it airborne again. And the larger the fish, the higher the glide. If caught by a gust of wind, a flying fish might be swept 20 ft (6.1 m) into the air and land on the deck of a passing ship.

Other fish also 'fly'. Flying gurnards have enormous, thin, scalloped pectoral

FLYING FROG *Wallace's flying frog can extend the membranes between its toes and glide – with little aeronautical skill – through the air.*

fins, which can be split into an upper wing-like part and a lower undercarriage of movable fin rays. The most dangerous fliers – dangerous to people, that is – are the needlefish. These are long, thin, sharp-beaked, fast-swimming relatives of flying fish. They break through the water surface, shoot 16 ft (4.8 m) into the air like javelins,

and cover a distance of 80 ft (24.4 m). They can also scull across the surface using their tail. Fishermen unfortunate enough to be in the way when needlefish jump have been speared and even killed when a fish has pierced a vital organ.

TREE JUMPING

Perched precariously on a branch, the Asiatic gliding frog shoots out powerful back legs, but instead of landing a few inches away on the next branch, it launches into the air and glides several yards to the next tree. As it flies through the air, it extends its arms and legs, bends its body into a concave shape and stretches the webbed skin between its elongated fingers and toes to form four small but effective parachutes. The membranes, normally used by its more amphibious relatives to help them swim more efficiently, enable this aerial amphibian to 'paraglide' about its tropical forest home. In a single flight it can cover up to 150 ft (45.7 m). It uses its aeronautical abilities to enlarge its feeding range without having to descend one tree, cross the forest floor where it would be vulnerable to snakes and other predators, and laboriously climb up the next bough.

Another species of frog from central America, however, has given us a new insight into why a frog should want to fly at all. It is a tree frog, *Agalynchis*, which, as its common name suggests, spends most of its time in the trees, but must lay its eggs in water. Come the rains, there is a mad dash for the ground, as the first frog to reach the new ponds formed by the rain will be the first to mate. The most effective way to get down is to fly, or rather to parachute. The frogs leap from the canopy with arms and legs outstretched and their webbed toes spread out, and glide at an angle, landing on the leaves of the nearest arum lily or palm to

MARSUPIAL GLIDER
The Australian sugar glider floats from tree to tree using flaps of skin between its front and rear legs.

absorb the shock of landing. The frogs – sometimes as many as 200 in a dense, writhing mass – get together in such large numbers in order to outwit predators such as monkeys, snakes and birds. So many eggs are produced at the same time that preda-

FAST RUNNERS

The fastest land animal is the cheetah. With a giant stride, claws like running shoes and a specially flexible spine, the cheetah can reach speeds of up to 60 mph (97 km/h). The fastest running bird is the ostrich, which can travel at 30 mph (48 km/h) over long distances for 15-20 minutes without slowing. Among the land invertebrates the predatory tiger beetle is the fastest runner on six legs. It has been clocked at a speed of 24 in/sec (61 cm/sec), or a little more than 1 mph (1.6 km/h). This may seem slow, but the $1/2$ in (1.3 cm) long insect is actually going at 40 body lengths per second, the equivalent of a racehorse travelling at 200 mph (320 km/h). The fastest land invertebrate, however, has eight legs, not six. It is the sun spider, a relative of the scorpions. It can chase prey at 10 mph (16 km/h).

tors are quickly satiated and the majority of the offspring survive.

Another extraordinary and unexpected aviator is the flying snake. At one time its behaviour was dismissed as a fanciful traveller's tale, but we now know that its flying ability is real. There are several species, and all are first-class tree-climbers, able to shin up rough bark with the help of special

broad scales on the underside of the body. When ready to change trees, a snake will speed along a branch and out over a clearing. As it begins to fall, it flattens its normally round body into a concave, ribbon-like channel and undulates in S-shaped coils. The snake is then able to glide, its shape enabling it to fly at an oblique angle to the ground. It has little manoeuvrability in the air, but by wriggling its body it is able to change its flight path and land close to the place it had intended to go.

The ability to fly or, more accurately, glide in mammals is more widespread than one might first imagine, and it has evolved independently in several species. Bats are the only mammals that lift themselves off the ground and actually fly, but there are also flying squirrels and the unrelated scaly-tailed squirrels, gliders, colugos and even a flying marsupial mouse.

The glider or flying possum is a marsupial or pouched mammal from Australia and New Guinea. It moves rapidly through open forest; one species, the sugar glider, is able to glide 165 ft (50 m) at a time. It has a thin, furred membrane that stretches from its front legs to its back legs on each side of the body, giving a rectangular-shaped flight area, and it maintains control with its tapering, thick-furred tail. There are several species. The largest, the greater glider, demonstrates the steepest glide (40°), but like the highly manoeuvrable smaller species it can make a midair right-angled turn, and can brake and tilt at the end of the glide to land on the target tree with all four feet with some degree of accuracy.

A glider completes its glide by swooping upwards onto a tree-trunk and lands with an audible 'plop', sometimes the only indication that gliders are about. Slow-motion photography has revealed that the glider's landing approach is not as sophisticated as one might think. It actually collides violently

LIVING HELICOPTERS The ruddy darter dragonfly (above) can hover, fly backwards and accelerate forwards to gather up flying insect prey. Right: The structure of the dragonfly wing resembles that of an early biplane, yet it evolved over 300 million years ago.

with the tree. Elongated fourth and fifth fingers tipped with extra-large claws enable it to hang on tight, however, preventing it from bouncing back on impact. When not using its flight membrane, or patagium, the glider simply holds it against the side of its body, where it is barely visible as a wavy fold of skin.

The colugo or flying lemur, also known as the 'skin-wing', is about the size of a domestic cat and resembles a kite when gliding. It has an even larger flight area than the glider – the most extensive of any flying mammal. The furred membranes stretch from the neck to the tips of the fingers and continue to the tip of the tail. With such a large 'wing' the colugo is a long-distance glider. Flights average more than 350 ft (107 m), although one individual has been

seen to glide 450 ft (137 m) through its South-east Asian rain-forest home and only lose about 40 ft (12 m) in height. Despite the name flying lemur, it is not a lemur at all. True lemurs are confined to Madagascar and none fly. Colugos live in Malaysia and the Philippines.

POWERED FLIGHT

Powered flight is reserved for the insects, birds and bats. Indeed, insects were probably the first animals to fly, and in the intervening 300-400 million years since they took their first flight they have evolved into remarkable flying machines that would be the envy of any aircraft designer.

Like miniature helicopters, dragonflies hover in front of prey, and hawk moths, with wings moving at 20-50 cycles per second, hover in front of flowers. The common housefly, no slouch when it comes to flying, can decelerate instantly from full flight speed, hover in one place, turn about in its own length, fly upside down, do a loop with its front feet extended, roll its body and land on a ceiling, and all in a fraction of a second. The secret to this manoeuvrability lies in the wing design.

Insect wings are remarkable pieces of engineering. A strong, flexible membrane is supported by ultra-lightweight longitudinal and cross veins. In some species the membrane is no more than $1/1000$ mm thick, yet it can stand up to the enormous forces generated as the insect flaps its wings. Some insects, such as the dragonflies, mayflies

NOISY FLAPPER *The 'clap-and-fling' method of flying can be heard quite clearly when a garden fantail pigeon takes off.*

and grasshoppers, have corrugations on the wings that give greater strength and rigidity without adding too much weight. The membrane of dragonfly wings also helps to give strength to the entire structure in what engineers would term a 'stretched skin', like the fabric skins of early biplanes. The wings taper, a feature that makes the ends less rigid, but insects tend to bump into things and so the wing tips are built to yield on impact and then quickly recover their shape. There are also rigid veins to keep the basic wing shape when it is flapped, and veins that deform like the flexible parts of children's drinking straws, to bend and change the shape of the aerofoil according to circumstance. There are also shock absorbers, counterweights, and ripstop mechanisms that prevent holes from tearing further.

These features are familiar to aircraft designers, but the insects (and birds) also have unconventional solutions to staying airborne. Butterflies, lacewings, bugs and several other insects share with birds the 'clap-and-fling' method of flying, in which the wings strike each other directly above the animal's body and then fling

NOISY FLAPPER *The 'clap-and-fling' method of flying can be heard quite clearly when a garden fantail pigeon takes off.*

rapidly apart. In birds such as the pigeon, there is a clearly audible noise at the point when the wings touch. During the first part of the 'fling', bird and butterfly wings remain in contact along their trailing edges, but they separate in a different way, depending on the species. Bird wings tend to break apart cleanly, whereas butterfly wings peel apart, starting with the leading edge. Air flows between each wing, creating additional lift at the start of the stroke. The technique has no parallel in man-made aerofoils.

These remarkable revelations are made possible by high-speed still and motion photography. In work that began in 1989, using laser beams to trigger a special high-speed shutter, Dr John Brackenbury of Cambridge University has been able to isolate actions that can only be seen in $1/10\,000$ of a second. Indeed, he has been able to show that some insects use these techniques in movements other than flying. Bush crickets, for example, not only open their wings briefly when they jump, but they also use the same peeling action on the clap-and-fling movements of their wings to boost the height of their enormous leap.

FROZEN IN FLIGHT *A strobe flashlight isolates a green lacewing taking off from a bud. The eight images are separated by intervals of 20 milliseconds.*

EATING OUT

4

NIGHT HUNTER *The barn owl swivels its head in order to get a better 'fix' on its prey.*

THERE IS AN ARMS WAR GOING ON CONTINUOUSLY IN THE WILD. PREDATORS AND PREY — THE FORMER PROGRAMMED TO SEEK AND FIND, THE LATTER HIGHLY TUNED TO DETECT AND ESCAPE — HAVE BEEN HONED TO PERFECTION DURING MILLIONS OF YEARS OF EVOLUTION. ONE MUST OUTDO THE OTHER, FOR BOTH MUST EAT AS EATING IS ESSENTIAL FOR LIFE. SOME ANIMALS EXCLUSIVELY GRAZE PLANTS; SOME CATCH AND CONSUME OTHER ANIMALS. AND THERE ARE THOSE THAT SCAVENGE ON THE SCRAPS THAT ARE LEFT OVER. EACH IN ITS OWN WAY MUST BE ABLE TO LOCATE ITS SOURCE OF FOOD ACCURATELY AND BE IN THE RIGHT PLACE AT THE RIGHT TIME TO CROP IT OR CATCH IT.

ON TARGET *A three-horned chameleon captures a cricket.*

FINDING FOOD

Predators use all their senses to locate prey, but hunters in different habitats tend to develop one or two senses more keenly than the others, the adaptation often determined by whether they hunt on land, at sea or in the air.

In the search for the next meal, predators bring to their aid a formidable armoury of search mechanisms, all five senses, and responses to vibrations and electrical impulses. But they do not have everything their own way; their prey will do all in its power to escape, and many potential catches have equally acute senses with which to detect approaching danger. The result is a sensory arms race in which predators, if they are to survive, must be one sensory step ahead of their prey, and vice versa.

Few predators have the benefit of well-stocked larders readily to hand and their sensory systems must be able to operate from a distance. Predatory animals must also be able to register tiny movements or changes in the disposition of their prey. These are often so slight that they cannot be detected by humans. Predators also often have what appears to be an automatic response to certain characteristics, such as movement, shape and behaviour – this is known as a 'search image'. This search image enables a predator to see well-camouflaged prey that it might otherwise overlook. The process is not easy to understand in animals, but observations of particular predators suggest that, once spotted, the prey leaves behind an 'image' of itself in the predator's mind. Toads, for

PLAINS DRIFTERS *Enormous herds of wildebeest migrate across the savannah in search of fresh grass.*

ELUSIVE MEAL *An adult wildebeest is unlikely to be brought down by a lone cheetah, but a group of cheetahs can overcome a yearling calf.*

example, normally eat live creatures, but will snatch at a piece of moss if they have been eating spiders recently. Likewise, they will try to eat long, thin objects, such as twigs, if they have previously been feasting on worms.

WAYS OF SEEING

Vision, as a primary means of finding prey, has advantages and disadvantages. The advantages include the ability to detect and locate prey immediately and accurately, but there are significant drawbacks. Prey can hide behind a tree or a rock, be lost in fog or a rainstorm, or fool a predator with suitable camouflage. Underwater or at night, eyesight has many limitations, although underwater creatures and nocturnal animals tend to have large and highly developed eyes to compensate. Diurnal predators (those that hunt during the day) such as birds – and humans for that matter – depend largely on vision for searching out food, although most animals see very differently from humans and their visual world is largely secret from us.

Birds of prey have big eyes, and the transparent cornea at the front is flat, unlike the curved cornea in humans, allowing a bird to hold more of its field of vision in focus at any one time. At the back of the eye, on the light-sensitive retina, there are between two and five times more light-receptor cells than there are in the human eye. This gives birds a high-definition view over a large area, a view that is

only available to us with technological aids.

A further refinement of the eye of birds of prey is in the fovea – a point on the retina in each eye where vision is sharpest. Eagles, hawks and falcons have two in each eye, and in a bird of prey the fovea produces a visual image about eight times as acute as our own.

A visible feature that distinguishes predatory birds from others is that their eyes are set not at the sides of the face, but the front, giving binocular vision and the ability not only to spot prey but also to locate it accurately by judging distance. The eyes of an eagle fit tightly into their sockets, and the bird is unable to glance up and down without moving its head. Owls have tubular-shaped eyes filling all the available head space. Like the eagle, instead of moving its eyes, an owl turns its head, sometimes rather comically, almost through 360 degrees. This also gives rise to the curious bobbing behaviour of

BIG EYES *Burrowing owls in North American use their large eyes not only to detect prey but also to look out for swooping predators.*

an owl when it spots something interesting in the distance and moves its head up and down or from side to side to assess how far away it is. Eagles have even been known to turn their heads upside down.

With such sophisticated eyes, a bird of prey is capable of some remarkable feats. The kestrel, for example, can spot the movement of a mouse in the grass from a height of 1 mile (1.6 km) above the ground. And a martial eagle, one of Africa's largest birds of prey, was once seen to take off from a hill 4 miles (6.4 km) from an unsuspecting guinea fowl, fly directly to it, swoop down, grab the bird and carry it away. A person

VISUAL PERCEPTION *The eyes of the red-tailed hawk of North America can see eight times as sharply as those of a human.*

armed with a pair of powerful binoculars could see a bird of the same size from about 1 mile (1.6 km) away.

The mammalian answer to the bird of prey's visual box of tricks is known as the 'visual streak'. Lions and other cats do not have a central fovea, as humans do, but instead a broadly horizontal band of ultrasensitive light receptors. This gives them sharp vision ahead and to either side, so that they are able to see the movement of prey on the ground rather than the relatively small disc of acute vision in the centre of our visual field. They are less interested than we are in events taking place above and below them, concentrating more on feeding opportunities that present themselves directly ahead. The cheetah takes this to an extreme, with a very narrow strip of light-sensitive nerve cells across the retina, enabling it to see prey clearly against its immediate horizon.

Smaller animals, such as the chameleon, also have well-developed eyes and rely on vision to track down prey. The chameleon has the particularly useful ability to move its eyes independently of each other and in all directions. When prey is located, however, both eyes are brought to bear on the target, giving this extraordinary reptile binocular vision and greater accuracy when its long tongue shoots out to capture a meal.

The acute vision enjoyed by hawks and eagles, cheetahs and chameleons is well-known. However, it is a tiny ¼ in (6 mm) long 'opossum' shrimp called *Dioptromysis*, living in the coral reefs of Belize, that has the world's smallest pair of binoculars. At first glance, the shrimp appears to have the stalked, compound eyes with 800-900 tiny lenses characteristic of crabs, lobsters and

INDEPENDENT EYES *With one eye looking backwards and the other forwards, the parson's chameleon watches out for both prey and predators.*

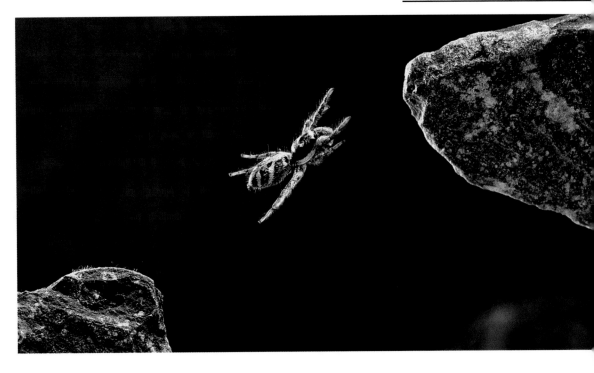

LONG JUMPER *The tiny jumping spider is able to cross a chasm or pounce on prey with a leap that would be the envy of an Olympic athlete.*

prawns, but closer inspection reveals a lens three times larger than the others in the middle of each eye, but pointing backwards. The lens is, in effect, an eye within an eye, with a narrow angle of vision of only 15-20 degrees and much higher resolution – six times – than the smaller facets. Like binoculars, the lens requires high light levels in order to function properly, and the shallow waters over white sand illuminated by a bright sun provide the ideal conditions.

Observations suggest that the backward-facing position is just a resting mode. Like somebody with binoculars hanging round their neck, the shrimp swivels its binocular eyes to the front when it has something specific to look at. What the shrimp finds interesting is not clear. Animals with a sensory system of this sort usually have food on their mind, but as yet, research has failed to identify the tiny shrimp's prey.

The dominant eye of jumping spiders, known for their ability to discriminate visually between prey and partner at a distance, contains a special pit, similar to the specialist fovea of eagles, and it enables them to see one-and-a-half times better than spiders with 'ordinary' eyes. It is a case of convergent evolution, in which nature has solved a problem in the same way for different

COLOUR VISION *Different animals see different parts of the visual spectrum, some seeing parts that others cannot.*

THE VISUAL SPECTRUM

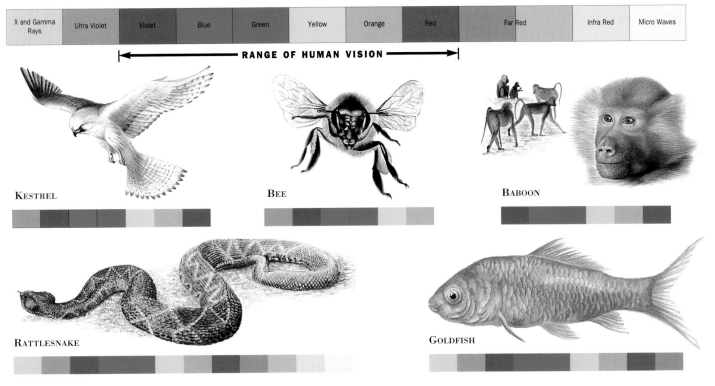

X and Gamma Rays	Ultra Violet	Violet	Blue	Green	Yellow	Orange	Red	Far Red	Infra Red	Micro Waves

◄———— **RANGE OF HUMAN VISION** ————►

KESTREL

BEE

BABOON

RATTLESNAKE

GOLDFISH

AERIAL PERCH *A kestrel hovers over its target, holding its position by flying gently into the breeze with its wings and large tail fanned out.*

groups of animals. The jumping spider's eyes have a corneal lens and a retina with four layers of light-receptive cells embedded in a matrix containing a special pit. The pit has a refractive surface – it focuses and concentrates the light – and this serves to increase the focal length of the eye (the distance at which it can focus objects).

Bushbabies have a cross-shaped pattern of densely packed light-receptive cells on their retina. These nocturnal creatures need to cock the head, like owls or eagles, in order to focus an important image, such as a juicy cockroach, on the sensitive cells – yet another example of convergent evolution.

Some animals not only see far more in a glance than we do, and more acutely, but they also see extra colours, which gives them a totally different view of the world. Most birds and mammals see what we see, as we know from experiments in which different species have been subjected to tests in which they are invited to recognise both shades of grey and different colours. However, at either end of the colour spectrum are colours – infrared and ultraviolet – not normally visible to humans, but insects and flowers have been found to communicate in the ultraviolet. Photographs taken using film sensitive to ultraviolet light have revealed that flowers have ultraviolet nectar guides, like runway lights on an airfield, that only insects and the occasional hummingbird species can discern.

Seeing in the ultraviolet is not confined to species interested in flowers. Kestrels also locate their food by an ability to detect ultraviolet light. Voles leave trails of urine and faeces for other voles to follow. Vole urine, however, contains chemicals that reflect ultraviolet light. They cannot see this, but

OBVIOUS TARGET *An infrared image of a mouse is superimposed on a picture of a mouse, simulating the 'view' a rattlesnake has of its prey.*

kestrels, in the course of their evolution, have developed the ability to see it. By watching for the crisscross of urine trails in a field, a hunting kestrel is able to see where vole populations are concentrated. It then hovers over these sites in the expectation of gaining a clearly signposted meal.

At the other end of the visible spectrum is the infrared, a wavelength of light that we 'feel' – as warmth – rather than see, and not surprisingly nature makes use of this too. Pit-vipers and rattlesnakes have pits on their snouts containing membranes with which they can detect heat from a living body. They can appreciate changes in temperature as minute as 0.005°C. In pitch darkness a western diamond-back rattlesnake can detect the presence of a mouse as long as the unfortunate rodent strays within 6 in (15 cm) of the snake's nose. Constricting snakes, such as boas and pythons, have similar receptors in their lips.

THE SMELLING AND THE SMELLED

A less precise but nevertheless effective way of finding a meal is to sniff the air. An odour is liable to disperse quickly and can be washed away by rain, but when it is available it can give away an animal's presence to a predator. Some predators can even obtain a fix on the prey's position by following a smell back to its source, as the closer the predator gets, the stronger the smell.

Although they breathe through nostrils, some snakes and their relatives the lizards find prey by 'tasting' smells in the air. Chemicals in the air are trapped on the moist tongue and pressed into the Jacobson's organ – a pair of pits lined with sensory cells – in the roof of the mouth. The sense of smell of narrow-tongued monitor lizards is better developed than that of their broad-tongued relatives. The 10 ft (3 m) long Komodo dragon, for example, which takes its name from an island in Indonesia, depends on an acute sense of smell for homing in on the odour of carrion.

The smell of a putrefying carcass may have every Komodo dragon in the district heading upwind, but not all scavengers are attracted by smell. Among the vultures, the majority use highly developed sight to spot

SEEING WITH SOUND

Why do bats hunt at night? About 50 million years ago the oldest known bats were competing for food with small hawks, falcons and birds related to modern-day rollers. To avoid being preyed on themselves, the bats flew at night. There was still danger from owls, but it has been calculated that, by flying in the dark, bats increased their chances of survival a hundredfold.

Bats are not reticent: they yell at their prey, emitting high-frequency sounds through either the mouth or nose, where curiously shaped nose-leaves focus and project the sounds as a beam. The beam bounces off the target and the bat registers the returning echo. This system of echolocation enables bats to find, chase and catch flying insects.

A small, insect-eating bat, such as a noctule, cruising about in a leisurely way, avoids bumping into anything by producing pulses of echolocating sound at the rate of about 10 pulses a second. As soon as it picks up an insect, the pulse rate of the beam rises rapidly to 50 beats a second. As the bat closes on its target, jigging this way and that, it needs more details about the insect's size, speed and direction, and produces a buzz

BIG EARS *The large ears and nose-leaf help the gothic bat detect its prey.*

of signals containing 200 or more pulses per second. To gain even more information, it can adjust the frequency of its sounds. Many small bats, known as FM (frequency modulation) bats, sweep their echolocation sounds downwards at the end of each pulse. In doing so, they increase the bandwidth of their signals and gain additional information about the target.

The greater horseshoe bat – one of the so-called Doppler bats – has a slightly different system, which relies not on the time taken for an echo to return, but on a comparison between the frequency of the outgoing signal and that of the returning signal. The bat knows what frequency it sent out, but the returning frequency will be different if the target is moving in relation to the bat – a phenomenon known as the Doppler Effect. If the target is moving away, the frequency of the returning signal drops, but if the prey is coming towards the bat, the frequency rises. The effect can be heard when a fire engine or

police car with its siren blaring passes by. The difference in frequency between the transmitted signal and the echo is a means of measuring the relative motion of predator and prey.

Doppler bats, such as the greater horseshoe, have a less frantic hunting tactic than their smaller cousins who use the FM system. The horseshoe might even remain stationary, sometimes at the roost site, rotating its head and scanning the surrounding area with a narrow

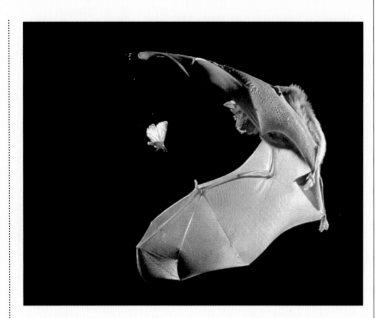

beam of sound, much like ground-to-air missile radar. When contact is made, the bat flies out, catches the prey, and returns to the launch site.

Some bats do not rely on the production of high-frequency sound signals to locate their prey. Instead, they listen for the sounds made by the prey. These bats are known as the 'gleaners', and a typical gleaner is the pallid bat of North America.

FLYING OUT *A 'Doppler' bat emits ultrasonic pulses; the echoes bouncing off an escaping insect are of a lower frequency than those sent out.*

MOTH CATCHER *The greater horseshoe bat launches its strike after scanning the dark using an echolocation system like ground-to-air radar.*

Moths and many other flying insects 'warm up' before take-off. They must raise the temperature of their flight muscles before they can fly efficiently, and the pallid bat homes in on the telltale wing vibrations.

The fringe-lipped bat of Central and South America navigates through the rain forest using normal high-frequency sonar, but when it detects the call of the tiny mud-puddle frog, which it prefers to jungle insects, it switches off its echolocation system and focuses on the lower-frequency calls of the frog. It swoops down from the forest canopy, grabs a frog and heads for home.

FLYING IN *If an insect flies towards the bat, the returning echoes are of a higher frequency than those that were emitted.*

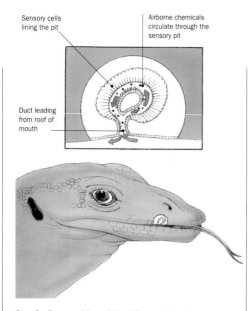

Sensory cells lining the pit

Airborne chemicals circulate through the sensory pit

Duct leading from roof of mouth

TASTING THE AIR *The lizard's tongue collects airborne odours and wipes them against the Jacobson's organ in the roof of the mouth.*

food. Some New World species, however, use smell to track down carcasses in the most inaccessible places, such as dense forests. The turkey vulture, a New World species more closely related to storks than

to birds of prey, is thought to have a well-developed sense of smell, an attribute that makes it a useful friend to the king vulture, which appears not to have this ability. King vultures string along with turkey vultures, the disturbance caused by the latter leading the former to the hidden site of the corpse.

Another unexpected scavenger flies among the trees of the New World forests in search of dead bodies. It is not a bird but a carrion bee. The discovery was made when an American researcher working in Panama put out the carcass of the Thanks-giving turkey for the local cats and attracted

a swarm of bees. These bees, which do not sting, had been observed around dead bodies before, but it was thought they were simply after the juices. Closer examination of their mouthparts, however, revealed that they have five large, pointed teeth on each mandible. With these they are able to cut into flesh, and when they find a carcass, like most social bees they work together. When a bee locates a carcass, it lays down a trail of pheromones, a chemical message that triggers a particular response in other members of the same species, back to the nest in order to recruit more help. As the bees

SCAVENGERS *White-backed vultures look for other vultures dropping down to feed. The knock-on effect can bring vultures from miles away.*

LITTLE SHOCKERS

The electrical function of the nerves and muscles has developed to such an extent in some fish that they can use the electrical field they produce to detect, debilitate and even kill prey.

A group of freshwater fishes, known as the 'weak' electric fishes, surround themselves with an electrical field that they can switch on and off at will. In the turbid waters in which they live, they can find their way about, communicate with each other and detect the presence of suitable items of food. The field is generated by modified muscles or nerves, depending on the species. Some species, like the knife fishes and elephant trunk fishes of South America and South Africa, have modified body and tail muscles.

Prey detection works very simply: when an object enters and interferes with the electrical field, it is picked up by sensory cells on the fish's body. Some fish, such as the knife fish, focus on the object by wrapping

their body around it, whereas elephant trunk fish move backwards and forwards, scanning the object and increasing their electrical pulse rate, in the same way that bats increase their rate of sound pulses when approaching an object. The smaller weak electric fishes tend to locate and eat small aquatic insect larvae and worms, but the larger specimens will attack other fish.

At the other end of the scale are the so-called 'strong' electric fishes. Their exploits are legendary, but the incredible strength of their electrical discharges is a matter of record. There are three notable 'shockers': the electric ray, the electric catfish and the electric eel. The electric ray or torpedo ray, an ocean fish, was known to the ancient Greeks as the narke. It grows to over 6 ft (1.8 m) long and stores electricity in two blocks of modified muscle on either side of its flattened body. Electrical discharges of 220 volts have been recorded – enough to make an

electric lightbulb glow. The fish cannot manage this for long, however: it must retire after the first few attacks to recuperate and, literally, recharge its batteries.

The freshwater electric catfish lives in Africa. It can grow to over 3 ft (0.9 m) in length and deliver a 350-450 volt blow. But the 'strongest' of the electric fishes must be the 9 ft (2.7 m) long electric eel that lives in holes and hollows in the banks of

South American rivers. It can generate a staggering 550 volts, not just once or twice, but several times a minute for as long as necessary. A single discharge is enough to knock out a human being.

POWER STATION *The torpedo or electric ray is able to detect and even stun fast-swimming prey with a powerful electrical discharge.*

WHITE DEATH *The great white shark uses sound, smell, feeling, sight and an electrical detection system to search out its prey.*

descend in large numbers they displace any other insect competitors, such as flies, and vigorously chase them away. They share this behaviour with mammalian scavengers, such as hyenas and lions, which dominate food sites by aggression and sheer weight of numbers. Having taken over, some members of the swarm form a circle and tear the skin. They make a small hole through which the rest of the bees can enter and methodically demolish the insides of the corpse. As they chew, they spread an enzyme over the meat and partially digest it before taking it back to the nest, where they regurgitate it for others.

Curiously, the bees seem to have an arrangement with a species of ant that also has an interest in the carcass and which the bees do not attack. The ants in turn do not harass the bees even though they will attack small wasps that may fly in. The ants may have arrived at the carcass first, but they will give way when the bees arrive, returning again at night. It is thought there could be a mutual understanding, a chemically induced compromise, between ant and bee, which has caused them to exercise a shift system – bees by day and ants at night. The most common feed available to carrion bees is the meat of amphibians or reptiles, although they do take an interest in much larger dead bodies. A dead frog can be reduced to its bare bones by a small swarm of 60-80 bees in about three hours, but a 1000-strong swarm may tackle a dead monkey or anteater and strip it bare in a few days.

THE REALM OF ALL THE SENSES

For some predators, the special development of just one or two senses is insufficient to take advantage of every opportunity and to take account of sudden, unpredictable changes in their environment and circumstances. These hunters need all their senses to be in a highly tuned state, and, like modern spacecraft, they have back-up systems that can drop into place as the situation develops.

The ultimate all-round predator – refined by millions of years of evolution into a sophisticated hunting and killing machine – must be the shark. These active hunters are aided in their task of seeking out and eating by the most diverse array of sensory mechanisms used by any predator. Alerted first to the low frequency (40Hz and below) sounds of a struggling fish from 1 mile (1.6 km) away, the shark can smell blood or body fluids at 1/2 mile (800 m). Taking a zigzag course in order to compare concentrations in different directions, it can detect one part of tuna extract in 25 million parts of water – the equivalent of ten drops in an Olympic-sized swimming pool.

If that's not enough, recent discoveries have revealed that some sharks also 'sniff' the air in their search for prey. The oceanic white-tip shark is a surface-dweller, patrolling the surface waters of the world's oceans in search of predatory opportunities. Its nostrils are at the front of its head, and when the white-tip raises its head out of the water it captures and retains air bubbles in its nostrils. The bubbles are broken up by the folded inner surface of the nostrils; any odours in the air spread through water in the nostrils, and the shark detects them more rapidly.

Below the surface, at 330 ft (100 m) away, the shark's lateral line can 'feel' the presence of the prey by detecting vibrations in the water; at 80 ft (24 m) and in almost total darkness, the shark can see the movements of its prey, and in colour. Its eyes are ten times more sensitive to dim light than ours are. If the shark is heading towards the surface, the pupils contract and the reflective tapetum

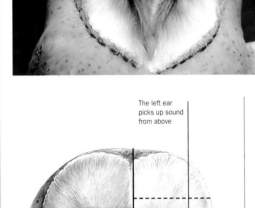

The left ear picks up sound from above

The right ear picks up sound from below

EYES AND EARS *The disc-like face of the barn owl is shaped to collect sights and sounds that enable it to hunt in the dark. The two halves of the disc focus sounds from different directions into the owl's ears.*

Infinitesimally small movements can be important in sound detection, too. The outsized external ears of the fennec fox, bat-eared fox and aardwolf gather in all the available sound. With these the animals can detect the minute scufflings of creatures as small as termites in their underground tunnels. Such animals hunt mainly at night, sometimes in an apparently erratic way. Bat-eared foxes from southern Africa, for example, race about their feeding areas, stopping momentarily to put an ear to the ground in order to pick up the scrabbling sounds of busy termites.

Sound detection is vital to some other night hunters as well. Owls, for example, are able to hear the faintest rustlings of a mouse in the grass, but they do not rely on large external ear flaps, like fennec foxes, for this; instead, the entire facial disc gathers in sounds. The heart-shaped face of the barn owl, for example, with its tightly packed feathers, resembles a pair of parabolic reflectors, focusing sounds on the ears themselves.

In the same way that eyes set in the front of the head make stereoscopic vision possible, the shape of the face and the position of the ears provide the bird with a stereophonic capability that enables it to pinpoint the source of sounds. The ears are set one slightly above the other on either side of the head, which gives the owl up-and-down as well as horizontal information. The result is an amazingly accurate system of prey location that works even in total darkness.

behind the retina is covered temporarily by a pigment, to adapt to the brightness.

During the final attack the eyes are protected and a remarkable electrical detection system in small pits in the snout takes over. With these, the shark can detect the electrical activity associated with a beating heart. The sensitivity is so great that it can detect a change in intensity of a $^5/_{1\,000\,000\,000}$ volt of electricity per cm, the equivalent of a flashlight battery creating a field between two electrodes thousands of miles apart. In this way a shark can locate prey such as a flatfish, even if it is invisible, buried in the sand. Tests in aquaria have shown that sharks can detect not only living flatfish but also hidden electrodes. When the electricity is switched on, even with extremely small currents flowing, the sharks home in unerringly on the right spot.

DESERT FOX *The fennec fox's huge ears pick up the sounds of moving locusts, small birds and jerboas, which it hunts by night.*

FEASTING AND FEEDING

It is one thing to locate food, but quite another to enjoy it. Animals have many adaptations that enable them to catch and eat what they have foraged, scavenged or hunted, including modified jaws, beaks, teeth, wings, feet and claws.

Dining out involves more than locating food. Having identified a meal, an animal must then actually catch and consume it. Prey tends not to give in too easily, so if it cannot be taken by ambush and surprise, a bout of stalking, chasing and catching will follow that tests a predator's ingenuity, strength, fitness and stamina. Both predators and plant-eaters shorten the odds by adapting parts of their anatomy to help them cope with the demands of obtaining food.

The bills, beaks and body-plans of birds are very obvious specialisations, and one glance at a bird can usually tell us what and how it eats. Sunbirds and hummingbirds have long, thin bills in order to extract nectar from delicate flowers, while spoonbills and ducks have broad bills that they use to dabble for delicacies in water. Hawks, falcons and eagles have bills that can slice and tear at a fresh kill, herons have long stabbing bills that they use to spear fish. Parrots and some types of finch have developed tough, sturdy beaks with which to crack nuts and seeds. Indeed, the study of the beaks of finches in the Galapagos Islands and honeycreepers in Hawaii has given scientists a view of evolution in action.

PATIENCE REWARDED *When a fish is within range, the great blue heron strikes rapidly, spearing the victim with its upper bill.*

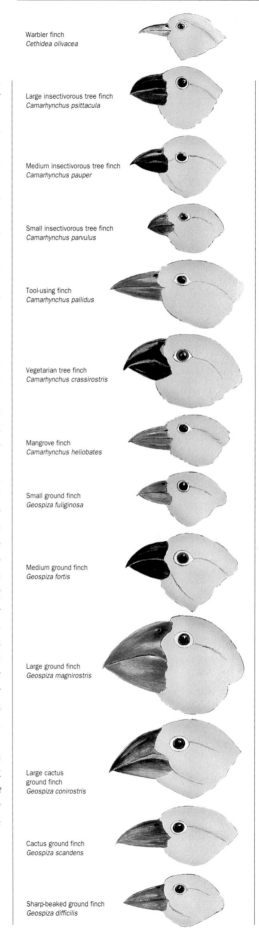

Warbler finch
Cethidea olivacea

Large insectivorous tree finch
Camarhynchus psittacula

Medium insectivorous tree finch
Camarhynchus pauper

Small insectivorous tree finch
Camarhynchus parvulus

Tool-using finch
Camarhynchus pallidus

Vegetarian tree finch
Camarhynchus crassirostris

Mangrove finch
Camarhynchus heliobates

Small ground finch
Geospiza fuliginosa

Medium ground finch
Geospiza fortis

Large ground finch
Geospiza magnirostris

Large cactus
ground finch
Geospiza conirostris

Cactus ground finch
Geospiza scandens

Sharp-beaked ground finch
Geospiza difficilis

VAMPIRE BIRD *The sharp-
beaked ground or vampire
finch has changed from
pecking at parasites to drinking
the blood of other birds. It is not
averse to pecking at the eggs of
boobies if they are left exposed.*

At some unknown point in the past, several individuals of a basic species of finch on a local migration on mainland Ecuador most probably overshot and ended up flying out into the Pacific Ocean. They landed exhausted on the Galapagos Islands, where most of them remained and bred. Gradually, as small, isolated populations adapted to local conditions and consumed different foods, the shapes of their bills changed. Those that had bills which could cope with local food sources survived and bred, while those that did not died out.

The finch is normally thought of as a seed-eating bird. But the sharp-beaked ground finch of Wolf Island, where water and food are scarce, has turned to parasitism in order to survive. It is believed to have acquired this habit while relieving its host, in the manner of African oxpeckers, of bothersome flies and lice, for which a pointed beak was needed. In this case the host is not a large, hoofed mammal, but a bird – the masked booby, a tropical relative of the gannet. The finch perches on the tail of this large sea bird and pecks at the rump, breaking the quills of the tail feathers and sipping the blood that is drawn as a result.

In the breeding season the finch waits until its host has left the nest and brood unattended, and helps itself to a large, white, oval egg. The finch and the egg are about the same weight. At some stage in its evolution, the finch must have recognised that the egg contained nutritious food and discovered how to break it open. Using its entire body as a lever, it flicks the egg against a rock, in the same way that a mongoose holds eggs between its hind legs and throws them against rocks in order to break them open and get at the contents. The finch may have to repeat this process many times before the egg cracks, or its irate owner returns. If it should succeed in cracking its prize before the mother booby reappears, the finch gorges on the yolk.

The 'vampire finch', as it is known, is just one of 13 new species of finch that Charles Darwin studied before he wrote *The Origin of Species*. It enabled him to study how, in a family of birds, changes in the

DARWIN'S FINCHES *From a
basic seed-eating bill, finches
on the Galapagos islands have
evolved different bill shapes to
suit a variety of tasks.*

HAMMERS AND CHISELS
A baby oystercatcher learns from a parent how to use its bill to hammer or lever open shellfish such as mussels.

environment or conditions – in this case, the need to utilise a new source of food – cause behavioural, physiological and anatomical changes to occur in a species. In short, the finches had to adapt or die.

Despite the obvious adaptations shown by their beaks, birds still have feeding secrets that are only just being discovered. How, for example, does an oystercatcher open a mussel, and how does a young turnstone learn its feeding technique? The bill of the oystercatcher – which has been closely observed in the south-west of England – is perfect for opening up tightly clamped prey. It is long and pointed and, viewed from the front, is taller than it is wide. It is just the right shape to prise open mussels, razor shells and scallops. But such a tool is no use unless the bird knows how to use it. In their first year of life, oystercatchers struggle with their food until they have learned the complex techniques for opening shells from older birds. Young birds are unaware which are the best mussels and

waste valuable time probing empty shells. Sometimes they get the shells stuck on their beaks and can no longer feed properly. Occasionally they scavenge on the leftovers from more experienced birds, and until they learn their trade this could be the last resort.

With a quick stab, the experienced oystercatcher can take a slightly open shellfish unawares and lever it apart with ease. This can also be done when the prey is closed, but may require more persistent jabs of the bill. If the shell has worn slightly thin in the face of the tides, the oystercatcher is able to hammer a hole in the shell without too much difficulty. The secret of success is to be able to recognise vulnerable shells, and this comes with experience. Birds are known to tap shells before selecting one that sounds flimsier than the rest. So much tapping and stabbing takes its toll on the narrow beak, but the tip regrows within two to three weeks.

Circumstances may require a species to be versatile in its feeding methods if it is to survive. Another shore bird, the turnstone, as its name suggests, turns over stones in order to find a meal, but this is not its only method of feeding. These busy waders, which follow the tide in and out, choose a feeding technique that is determined by the nature of the shore. On pebble beaches and in rockpools a bird will turn stones. Among piles of seaweed it will root about, bulldozing the algae to one side in order to reveal small molluscs and crustaceans. Along the water's edge the bird probes among the fronds of seaweeds, while in rocky areas it will hammer at barnacles. On the sandy beach, the turnstone pecks at debris on the surface, and on muddy estuaries it will dig down to find worms.

Female turnstones turn more stones than males do and are more skilled at catching the exposed prey. Each bird in a flock has its rank, and high-ranking individuals get first stab at the piles of seaweed, where food is concentrated. Lesser-ranking

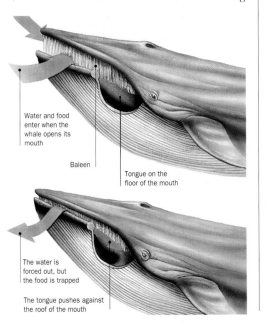

Water and food enter when the whale opens its mouth

Baleen

Tongue on the floor of the mouth

The water is forced out, but the food is trapped

The tongue pushes against the roof of the mouth

birds, excluded from exploiting the seaweed, must take to stone-turning. Males are dominant over females, and females must therefore become accomplished stone-turners in order to survive.

Whether a bird roots or not is also determined by the time at which it arrives at its winter feeding sites for the first time. Juveniles arriving early tend to become rooters, while late arrivals are forced into stone-turning, having been chased away by the rooters. And once a rooter, always a rooter. The way in which a turnstone feeds during its first winter therefore determines the way it will feed for the rest of its life.

There are additional benefits to being a rooter. With so much food in one place, a bird need not feed for as long as its non-rooting cousins so it has more time to be vigilant. Birds rooting on seaweed also tend to be further from the top of the beach where sparrowhawks or other predators lurk. The death rate for rooters is half that of non-rooters.

FOOD FROM THE SEA

The task of extracting a meal from the sea can be tackled in a multitude of ways. Filter-feeders sift sea water and filter out plankton and other small creatures: giant manta rays turn somersaults to scoop up water in the mouths on the undersides of their bodies; basking sharks skim the sea's surface; megamouth sharks patrol the midwaters, attracting a soup of deep-sea shrimps with a luminous mouth; and along the rocky shore, barnacles stand on their heads and scoop in food with their feet.

Whale sharks – the largest fish in the sea

MOUTHFUL OF WATER *Baleen whales gulp in a soup of krill or small fish, force the water through their baleen plates by pushing their tongue against the roof of the mouth, and swallow the food left behind.*

GAPING MAW *The enormous whale shark, which grows to 40 ft (12 m) long, skims the surface waters for the tiny crustaceans on which it feeds.*

– are also filter-feeders. They congregate in large numbers off the coast of Western Australia in the southern autumn. During March and April, every coral polyp along the entire Ningaloo Reef spawns at the same time. In an underwater 'snowstorm' male and female sex cells are released. Some combine, drift and then settle as new coral, but the glut of food attracts the next layer in the feeding chain – the zooplankton. It is this sudden superabundance of small crustaceans, or krill, and fish following the mass spawning of the coral that draws the whale sharks. They feed night and day, ploughing through the surface waters with mouths agape.

Each shark, up to 40 ft (12 m) in length, skims the surface. The krill have a daily vertical migration. They normally head towards the surface at night, where they feed on surface phytoplankton (the floating plant plankton), and hide in the depths by day. But during the days of plenty when the coral spawn, krill and sharks are active day and night. Joining them are giant manta rays, ploughing 50 abreast through the krill swarms. Juvenile reef fish take advantage of the sudden abundance of food too, and there are shoals of anchovies. The smaller fish, in turn, attract larger predators. Tuna

NIGHT FISHING *The fishing bat of Belize spots ripples in the water using its sonar, and hooks out a fish with the long claws on its feet.*

herd the anchovies into dense groups, which also attract diving shearwaters from above and huge-mouthed Brydes whales from below.

Fishing is an activity enjoyed and exploited by a diverse and motley group of animals. There are fishing bats with long claws on their feet to hook out their catch, skimmers with extraordinarily elongated lower bills, with which they 'skim' the surface, and ospreys with rough spicules (small needle-like structures) on their feet to deal with their slippery prey. There are fishing cats and fishing spiders. Brown bears hook salmon out of the river with their claws, and bald eagles feast on the salmons' spent bodies after spawning is complete. Cormorants chase fish below the surface and have been seen at depths of about 660 ft (200 m), while gannets, kingfishers and brown pelicans plunge dramatically from the air.

White pelicans, swimming in great horseshoe-shaped formations on the surface, drive fish into the shallows, a behaviour they share with dolphins and killer whales. Marlin and other bill-fish herd their prey into tight groups or 'bait balls' before they rip through the shoal. They are the fastest killers in the sea.

SPEED KILLS

On land and in the air, too, speed almost always helps to make a catch. Cheetahs, with a maximum speed of 60 mph (97 km/h), and peregrines, with an estimated stooping speed of more than 112 mph (180 km/h), are the champions. But the cheetah is only able to maintain its sprint for a short time, and having caught its prey, which it brings to a halt by swiping its legs away and tripping it up from behind, it is totally exhausted and in danger of overheating. It must pant for some time and catch its breath before carrying the food to a safe place. During those vital minutes the cheetah is vulnerable. Hyenas, leopards and lions may steal a cheetah's kill. The cheetah, which has speed but not bulk, is no match for a hungry scavenger.

Among birds, the roadrunner of the arid desert chaparral of south-west North America is also a speedy sprinter. It is able to fly, but prefers to chase small, speedy prey, such as lizards, on the ground. Paced by a car, the roadrunner has clocked up speeds of over 26 mph (40 km/h) – the equivalent of a 2.3 minute mile. Insects, too, use speed to attack. Europe's tiger beetle, which runs down fast-running insects such as ants, is the swiftest animal on six legs. It is capable of bursts of speed of up to 24 in/sec (60 cm/sec) or just over 1 mph (1.6 km/h), which may not seem very fast, but this $^1/_2$ in (15 mm) long insect is travelling at over 40 body lengths per second, the equivalent of a racehorse running at 200 mph (320 km/h).

The fastest invertebrate, however, has eight legs, six of which are used for running: the sun or wind spiders of arid lands are related to the scorpions and are particularly aggressive predators, using fang-like pincers to crush their prey. During the pursuit of darkling beetles and small lizards, the sun spider picks up the vibrations of its prey by means of sensory organs on the fourth pair of legs, which are held out in front like feelers. At top speed the sun spider is able to travel at 10 mph (16.1 km/h).

FISHING STRIPES *The body of the marlin changes from a uniform silvery sheen to blue and silver stripes when it attacks a shoal of small fish.*

BIRD RACER *The North American roadrunner chases its prey, often lizards, on the ground, giving credibility to its cartoon counterpart.*

Flying insect predators must also have speed on their side. Dragonflies are among the fastest – they are the hawks of the insect world, with flying speeds estimated at around 36 mph (58 km/h). Indeed, one group has been named the 'hawker' dragonflies, and like their relatives the 'darters' they have large, effective compound eyes that can spot prey from 33 ft (10 m) away.

Hawker dragonflies attack on the wing, while the darters fly out from resting sites like missiles from a launch pad. Both catch prey with their legs. The thorax, the middle section of the body, to which the legs are attached, is angled forward. This enables the dragonfly to form a spiny basket with its legs and scoop up prey in midair. It then takes the prey back to a perch and consumes it. Both types of dragonfly have their own, well-guarded territories, a behaviour they share with robber flies. These small predators also patrol a regular beat, their acute vision and rapid flight enabling them to outfly many other insects. The fly's mouth-parts resemble sharp, hollow spears with which they can pierce an insect's exoskeleton in order to suck out the contents, until nothing is left of their victim but the 'skin'.

COME INTO MY PARLOUR

Another creature famed for its ability to suck a fly dry is the spider. Most spiders rely on their webs to catch and consume prey, but this is not the only way. The use of disguise, glue and poison is also common. Some spiders, such as wolf spiders, depend on speed to catch a meal, but others show more ingenuity.

Jumping spiders can identify a potential victim up to 12 in (30 cm) away and pounce on their prey. The leap can be over 40 times the spider's own body length, and a safety dragline is spun as it jumps in case the spider should land short of its target and fall. However, some jumping spiders fail to live up to their name and are more devious instead. Some tropical species sneak up to the webs of orb-weaving spiders,

VEGETARIAN SPIDERS

Although spiders feed mainly on insects and other spiders, a few eat other foods. Spiders have a habit of dismantling their webs by eating them – it is a way of recycling the materials and rebuilding damaged webs. Young orb-web spiders, such as juvenile garden spiders, gain additional nutrients from pollen trapped on the sticky threads of the web when they are reeled in for recycling. Male crab spiders rarely capture prey, searching for and guarding potential mates instead. They gain energy and water from the nectaries of flowers.

drum on the silk to mimic insects and then jump on the resident spider as it rushes out. An Australian species, *Portia fimbriata*, has even perfected the art of catching other jumping spiders. It creeps up on a victim while its back is turned, moving at no more than 1/32 in (1 mm) at a time. If the prey should turn to face the stalker, the latter freezes, drawing in its palps (leg-like sensory appendages near the mouth) and looking to all the world (including scientists who have studied the spider) like a dead leaf – a behaviour known as cryptic stalking. When the prey is within range, *Portia* grabs it from behind and stabs it in the pedicel, the narrow part between the head and abdomen.

The tiny crab spider sits in flowers and mimics their colours and patterns. Motionless and perfectly disguised, it waits for insects to come along and feed. When a

DESERT JAWS *The sun spider has the largest jaws in relation to body size of any animal, but its most formidable weapon is its speed.*

hoverfly or butterfly alights, the spider grabs its head and bites into the victim's mouth-parts, extracting its body fluids. The crab spider's venom is so powerful that it can also tackle larger insects such as bumblebees and wasps.

Disguise is taken to an extreme by the thomasid spider of Australia, which exploits its likeness to the very species it most commonly preys on – the northern green tree ant. It waves its front two legs in the air like ant antennae; it also has two spots on the rear of its body resembling ant eyes. Whichever way it faces, the thomasid spider looks like an ant. It places itself on a branch frequented by its prey, but avoids contact with it. After a time it begins to stalk. Finally it pounces on an ant from behind and stabs

WEB WARNING *The black-and-yellow* Argiope *spider stores its insect prey in a silken coccoon; the crisscross pattern in the web acts as a warning to birds.*

it in the 'neck', injecting a poison. In order to avoid the jaws of the northern green ant, the spider retreats and waits for it to die. It then drags the corpse to the underside of a leaf, where it can consume its meal at leisure. If things do not go to plan and the ant attacks the spider, it puts out a safety line and lowers itself to the ground.

The trap-door spider conceals itself in a hole with a lid as it waits for its prey. It uses comb-like mouth-parts to dig a vertical shaft up to 8 in (20 cm) deep. The walls are waterproofed with saliva and earth and then lined with silk. A tight-fitting, hinged cap with bevelled edges completes the lair, and the spider sits inside. It can close the lid against any intruders, but usually leaves it

FALSE ANT *This North American jumping spider fools ants into thinking that it is one of them, and then grabs one and sucks it dry.*

slightly ajar and waits for passing insects. When the victim comes near enough, the spider darts out, taking care to prop open the door of the lair with its back legs for a quick retreat, and grabs the insect. The spider stabs its prey with its fangs, injecting poison together with enzymes that break down the victim's body tissue, and it then sucks it dry.

The most obvious food-catching technique used by spiders, however, is shown by the magnificent orb-weaving spiders. Webs spun by spiders today are the product of 180 million years of evolution, and they are very sophisticated pieces of animal engineering. An orb-web spider's web is capable of trapping large, bulky, fast-flying insects. The web is a framework of dry, radial threads surrounded on the surface by a single sticky spiral, and has enough give in the construction to absorb the impact of an insect without breaking.

In order for a web to work, it must be difficult for the prey to spot, but the sticky droplets of fluid that coat the spiral thread – the coating is roughly twice the width of the core – reflect more light than the silk used in the rest of the web. A spider in an open habitat must therefore produce a web with fewer sticky droplets, and risk losing prey such as moths and butterflies that can slip free of a web by shedding the scales on their wings. A spider in a darker environment, or a nocturnal, can decorate its web with more of the sticky fluid and catch almost any insect that blunders into it.

In order to deter something very large, such as a bird, which would tear through the silk no matter how tough and pliable it is, the spider builds in the equivalent of warning lights. To do this, many orb-web spiders weave conspicuous bands of loosely spun silk into the web. These strips are known as 'stabilimenta', and it seems that

they alert birds to webs in their flight paths. Webs without stabilimenta are often destroyed in springtime, when birds are particularly active.

GETTING THE BETTER OF PLANTS

A major problem for some herbivores is not to outwit prey, but to overcome the extraordinary defence systems of plants. Plants have been at war with animals for millions of years, and some animals have gone to great lengths to overcome their defences. Gorillas have found a way to deal with stinging nettles, elephants and giraffes are able to eat thorny acacia, and uakari monkeys crack open very hard nuts.

The uakari monkeys differ from all other New World monkeys in not having a long tail, but the most dramatic features are their large jaw muscles and sharp canine teeth. The combination enables the golden-backed species to crack open fruits with $^3/_4$ in (1.9 cm) thick shells, which is the equivalent of a human trying to bite into a coconut. The monkey goes to all this trouble in order to get at fruits that contain a minimum of poison. Softer fruits tend to have toxins that deter most seed predators from eating them.

The chimpanzee of West Africa is another nutcracker. For about two-and-a-half hours each day a group will sit about and crack the exceedingly hard nuts of the Coula tree and those of *Panda oleosa*. In order to crack them open, chimps learn to use a hammer and anvil – an example of the use of tools in the animal world. A branch or stone with a suitable recess, in which the nut can sit without slipping, is used for an anvil, while a branch or stone weighing up to 44 lb (20 kg) serves as a hammer. Babies learn this skill from their mothers, picking up titbits and even borrowing their mother's hammer to get started.

Leaves are less nutritious than nuts, but even so both birds and animals take advantage of them. The South American hoatzin (related to game birds), for example, has a gut like that of a ruminant with which to digest green leaves. Unlike a cow, which has an enlarged hind-gut where bacteria break down the tough plant-cell walls, the hoatzin has an enlarged foregut, much like a kangaroo. On its slow passage through the gut – the slowest of any known bird – the food is broken down by bacteria. These detoxify poisons before they are absorbed and release nutrients, but they also make vitamins, minerals and amino acids available to the bird.

It is the high level of protein-building amino acids in leaves that makes them attractive to an unlikely leaf-eater, the flying fox or fruit bat. As its name suggests, the fruit bat eats mainly fruits and blossoms, but fruits are low in proteins. The black fruit bat of Australia gains its essential proteins from

---

VAMPIRES SHARE BLOOD

Some bats have a taste for blood. Two kinds prefer the blood of birds – one species creeps up on chickens and bites them on the ankle – while a third species has a fondness for mammal blood. All species are a nuisance to livestock farmers in Central and South America, and they often visit the same creature night after night until it becomes too weak to stand. They do sometimes bite people: a leg, cheek, nose or forearm protruding from under the bedclothes can be inviting. And they can spread rabies. But, contrary to popular belief, a vampire bat does not have large fangs that leave puncture marks on its prey; instead, it has razor-sharp front teeth that take off a thin slice of skin without hurting or even disturbing the victim. It laps at the blood flowing from the wound with a grooved tongue, while clotting is prevented by anticoagulants in its saliva.

On returning to their cave, female vampire bats cluster together, grooming and licking each other. It seems to be a way of checking out who is present in the roost. Each bat must consume 50 per cent of its weight in blood each night to maintain itself. Two or three consecutive nights without a blood meal could mean death for a vampire. So, on returning to the roost site, the less fortunate vampires who have not come back with a full stomach – often youngsters learning their trade – beg from the well-fed and are given regurgitated blood.

The following night the roles may be reversed and the favour returned. It is one of the few examples of reciprocation in the animal world, and a behaviour pattern that ensures that many vampire bats will survive to the ripe old age of 18.

FACE OF A VAMPIRE *The razor-sharp front incisor teeth enable the vampire bat to slice a thin slither of skin from its victims.*

NUT-CRACKERS *The black uakari has teeth whose shape is specially adapted for its diet of nuts, fruit and vegetation.*

chewing leaves. It does not actually eat them, but simply chews them to release the nutrients and then spits out the masticated residue. This chew-and-spit method of feeding seems to be an adaptation for flight. A stomach filled with half-digested leaves would slow down a flying fruit bat, making it vulnerable to aerial predators such as hawks and eagles.

Gorillas, on the other hand, eat mainly leaves and stems and spend most of their time on the ground, or so it was thought until western lowland gorillas in the rainforests of Gabon were seen sitting high in trees munching fruit. But the most surprising fruit-eater comes from the South American rain forest. It is a tree frog. The frog makes its home in the small ponds that form in bromeliad leaves. It eats insects and other small creatures, but it also eats fruit – often on climbing plants.

Nectar is another plant product consumed by animals. Flowering plants use nectar as a means of attracting pollinators, and in doing so attract a motley group of 'sweet-toothed' creatures. Butterflies, moths and bees are well-known nectar-feeders, as are hummingbirds from the Americas and sunbirds in Africa. In Australia, the greyheaded fruit bats visit eucalyptus flowers and sip nectar, while in South-east Asia the

dawn bat is a frequent visitor to the durian tree. The durian explodes into flower periodically, attracting flocks of dawn bats that have been searching over a 25 sq mile (40 km²) area of forest on the lookout for the sudden superabundance of nectar. The passionflower is more dependable, producing one bloom per branch per night. Pallas' long-tongued bat, which feeds on passionflower nectar, must patrol a regular nocturnal beat in order to find the flowers as they come into bloom.

Bats cannot hover as hummingbirds can, and so they must snatch a meal during the fraction of a second in which they fly past. In flocks of at least 25 individuals, the Mexican long-tongued bat migrates north each summer to the Sonoran Desert of Arizona to feast on the flowers of the giant saguaro cactus and the agave. With a heart rate of 700 beats per minute, these small bats require large amounts of energy to survive. The flock circles a plant, each bat taking it in turns to swoop down on the flowers until the nectaries are exhausted. One of the bats – usually the last to visit the flowers – moves on to the next plant, and the other bats follow. The long tongue of the nectar-feeding bat sometimes measures up to one-third of its total body length. The length is achieved by pumping blood to the tip of the tongue, which causes it to extend in much the same way that blowing air into a coiled party whistle causes it to unroll.

Nectar-feeding may seem a relatively safe and harmless way of obtaining a meal, but tropical honeybees have found that there can be problems. In tropical areas high summer temperatures may cause the nectar to ferment, and up to 10 per cent of the volume could be alcohol. The nectar is taken back to the colony, where worker

bees turn it into honey, the vital food supply for the colony, and alcoholic nectar can result in 'spiked' honey. Similarly, if the bees cannot keep the temperature and humidity of the hive under control, stored honey may ferment.

The bees themselves can imbibe too much alcohol and become drunk, with disastrous consequences. Inebriated bees have difficulty flying, navigating and generally coordinating their movements. They smash into walls and trees, or just fall down and cannot get back into the air. If they should make it back to the hive they may miss the entrance or overshoot the landing area

FLOWER POWER

Many lizards eat insects, worms and other invertebrates, and some eat flowers and fruits, but a lizard that feeds on nectar was discovered in 1991. The Madeira lizard, a subspecies of the well-known wall lizard that lives on the isolated Atlantic island of Madeira, regularly visits several species of plants and licks nectar from their flowers. On the neighbouring island of Port Santo the same subspecies does not seek nectar. On the Caribbean island of Bonaire, juvenile blue-tailed racerunners also show the same behaviour, although the older lizards do not. Only young lizards, no more than 4³/4 in (12 cm) long, are small enough to push their heads into the flowers of the Turk's cap cactus, which they share with ruby-topaz hummingbirds.

completely. Their behaviour can be so erratic that they are denied access to the hive by the worker bees guarding the entrance. Drunken bees are then vulnerable to predators such as wasps, or they may die of cold. If many bees have succumbed to alcohol, the entire colony can suffer as honey production drops.

Fermenting fruit can have a similar effect on much larger animals, and it has not been unknown for drunken elephants to go on the rampage with dire consequences for any tree or bush – or village – that stands in their way.

AVOIDING BEING CAUGHT

Animals finding themselves considered as food try a variety of ways of making themselves less vulnerable, by being fast, less obvious, larger than life or too unpalatable to predators, either outrunning or outfoxing their pursuers.

The victims of predation do not take their fate lying down. In the evolutionary arms race they try to keep one step ahead of predators, and judging by the way in which prey animals are so numerous, they must succeed. If predators had things all their own way, the prey would be wiped out and the predators would quickly follow. Therefore an equilibrium exists between predators and prey. The vulnerable, however, have developed defence systems, either chemical or physical adaptations that momentarily (on a geological timescale) give them some advantage, and if they do not possess the wherewithal themselves they borrow it from some other creature.

One way to avoid being on a predator's menu is simply to disappear, and many creatures employ shading and countershading to do just this. The soft, pastel shades of a gazelle's body – dark above and pale below – make it almost invisible on the hot grasslands. The subtle pattern softens the boundary between the animal and its shadow.

Viewed in daylight, with the light coming from overhead, the animal's solid appearance is transformed into that of a uniformly shaded object. Fish employ a similar countershading. They are dark above and light below. Grey and black stripes on the back of a mackerel mimic the patterns made by scattered light as it passes through disturbed water. Seen from above, the fish blends in with its darker background. From below, its pale underside is lost against the sea's brightly lit surface. In the darker, twilight world of the ocean's mid-waters, where only small amounts of light penetrate, the same countershading is achieved with rows of luminescent light organs along the underside of the body. The midshipman fish (genus *Porichthys*) is lit up in

FOOD FOR MANY *The colouring of Thompson's gazelles (left), the prey animals of the savannah, helps them blend into the background. Above: The willow grouse matches its surroundings in summer; its plumage turns white in winter.*

BARK AND FEATHERS *A pair of Australian frogmouths mimic the boughs of the tree in which predatory birds have hidden.*

this way, with over 800 photophores, or light organs, along the sides and underside of its body, like an ocean liner at night. The glow blends in with the twilight background field. If the match is complete, the fish disappears from the view of any predator below.

On land in high latitudes, the trend is to wear white in winter and mottled brown in summer. For one creature a camouflaged coat is only a first stage to invisibility: in the face of danger, the willow grouse shuts down its body. Its heartbeat is reduced to a mere tickover, from 150 beats per minute at rest to 20 beats, and its breathing rate drops by 70 per cent. All signs of life are at a mini-

mum. If, however, the bird is found – by a wandering Arctic fox, for instance – in just one second the heart rate accelerates from 20 beats to 600 and the bird clatters into the air.

Horned frogs from South-east Asia resemble leaf litter on the forest floor; frogmouths closely mimic branches, including peeling bark; stick insects and stick caterpil-

lars resemble sticks; leaf insects look like leaves even to the extent of having veins, withered parts and blotches; and the young alder moth caterpillar resembles a bird dropping. But one of the most remarkable and fastidious mimics is the caterpillar of the dagger moth. The caterpillar itself has a greenish hue and is covered with a dense coat of fine translucent hairs. It has the overall appearance of the protective foam nest of the spittlebug or froghopper, which insectivorous birds tend to avoid. To further discourage

continued on page 122

UNPLEASANT TRICK *The giant swallowtail caterpillar spends part of its life camouflaged as a bird dropping. Overleaf: A well-camouflaged timber wolf.*

unwanted attention, it clears away any evidence that it has been in the area. Caterpillar damage on the leaves of a tree indicate to a passing bird that a juicy meal might be present and that it might be worth investigating. The dagger moth caterpillar, however, overcomes this problem in a rather clever way. Having finished its meal it backs

MONSTER MOTH *The caterpillar of the Indian lobster moth looks bigger than it really is and escapes being eaten.*

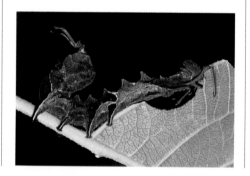

CANINE TRESSES
The megalipid caterpillar of Costa Rica has defensive hairs all over its body, and looks more like a toy dog than an insect.

down the leaf stem and bites off the leaf remnant, which then drops away, thus removing the evidence that there is a caterpillar feeding there.

DETACHABLE PARTS

While these careful eaters minimise predation by being inconspicuous, the hairy and spined caterpillars are often brightly coloured and munch their way through leaves in such a haphazard fashion that their presence in trees and bushes can be seen from some distance away. Their protection comes from poisonous hairs and spines, of which the emperor moth caterpillar has one of the most elaborate arrays in the insect world. The world's hairiest insect, however, must surely be a moth caterpillar from Central America belonging to the family Megalopygidae, which more closely resembles a miniature Pekingese dog than a moth larva. Its hairs detach and lodge in the throat of a predator, while spines below the thick thatch can deliver a painful sting.

Hairs, or setae, are also evident in spiders. They are not hairs in the mammalian sense, for a spider's setae are hair-like extensions of its external skeleton or cuticle. They may be organs of touch, taste or hearing, but only in the tarantulas, such as the 4oz (115 g) *Theraphosa* – the largest spider in the world with a leg span of 10 in (25 cm) – are they used in defence. The hairs on the pedipalps (mouth-parts on either side of the mouth) and the first pair of legs are rubbed together to make a hissing sound – the first line of defence. In addition, the hairs can be irritating to persistent predators as they

HAIRY SPIDER *A tarantula from the Amazon rain forest has a covering of irritating hairs, enabling it to eat its meal of a gecko without interruption.*

have barbs that cause burning and itching. In Arizona, for example, the hairs on a tarantula's legs seem to be specially adapted to inflame the nasal passages of one of the spider's main enemies, the grasshopper mouse. And if this fails to deter an attacker, the tarantula will bite with its large fangs, but it does this only as a last resort.

Web spiders, such as the orb-weaving argiope spider, tend to hide behind foliage at the side of the web when danger threatens, but if they are caught unawares they have their own escape mechanism. Scientists at Cornell University in the USA watched the reaction of a web-building spider when a stinging ambush bug landed in its web. It ran across the web with the intention of attacking the bug, injecting it with venom and wrapping it up in silk, but suddenly the bug turned on the spider and stung it in the leg. The spider froze, then jettisoned its leg – a defence strategy it shares with crabs, lobsters and lizards, which all escape predators by shedding trapped limbs or tails.

EARLY-WARNING SYSTEMS

Advance warnings of danger enable an animal to escape with a margin of safety. In order to do so, the animal's sensory system must be tuned accurately into the channel that will be most likely to warn it of danger, preferably at a distance. Small arthropods, such as ants, rely on vibrations to sound the alarm. For other creatures it might be sound, smell or vision. The freshwater fathead minnow of North America, for example, is able to detect the smell of a northern

pike's last meal, which tells it that a predator is about. The substance detected by the little fish is thought to be an alarm pheromone released by the pike's prey. It appears in the predator's faeces and then disperses in the water, where it is detected by other small fishes. After exposure to the substance, a minnow becomes more wary.

In the sea, Atlantic cod have their own early-warning system, an ability that appears to have developed very recently. Fishermen have noticed that some fish react to the signals from their echo sounders, and in laboratory tests cod have been seen to respond to bursts of sonar signals similar in frequency and pattern to the echolocating sounds emitted by toothed whales and dolphins. There is now speculation that we are witnessing the next evolutionary stage in the arms race between dolphins and their prey, and fish are about to take the lead: they can actually hear the dolphins approaching and take evasive action.

Monkeys receive early notice of danger from other members of their troop. Vervet monkeys, for example, have an alarm call that is not only specific to particular predators but also indicates the appropriate escape procedure. An approaching leopard

INFLATIONARY TACTICS

A successful strategy for defence on the part of some animals is to intimidate an attacker by looking much bigger than they really are. Inflation is the most energy-efficient way to increase one's size.

When a European common toad is confronted by a grass snake, it raises itself on stiff legs in order to appear taller and blows up its body to appear larger. As often as not, the snake ignores the curious creature expanding in front of it and slithers away. The toad is safe, for the time being. Several marine animals have found a similar way to survive.

The puffer fish and porcupine fish, the one with small spines and the other with large ones, can inflate their bodies rapidly with air or water, or both, and discourage an attacker such as a moray eel. The eel, even with its small, sharp teeth, is unable to grip the spherical body. When the danger has passed, an inflated fish deflates just as rapidly.

As recently as 1992, the crew of a French deep-sea submersible diving to 9850 ft (3000 m) below the south-western Pacific, off New Caledonia, discovered another animal that adopts the shape of a balloon in order to discourage deep-sea predators. The creature in question is an octopus, possibly of the genus *Cirrothauma*, which drifts close to the deep ocean floor. It is about 3 ft (1 m) across and resembles an inverted umbrella with paddle-like fins and a larger than average web connecting the arms. When threatened, the octopus changes shape, in much the same way as an umbrella is blown inside out by the wind, and swells up until it resembles a giant pumpkin. It is thought that this is sufficient to fool a deep-sea shark or other predators into thinking that the watery balloon-like object is not worth eating.

INFLATABLE TOADS *Confronted by an amphibian-eating snake, a Costa Rican toad stands tall and blows up its body in order to look bigger than it really is.*

CHORUS LINE *Birds in tight flocks avoid bumping into each other by watching the movements of their neighbours.*

triggers the 'spotted cat' alarm, which sees the entire troop heading for the topmost branches of the trees, where a leopard cannot climb. An 'eagle' alarm encourages all the monkeys to head for dense bushes and stay clear of the tree tops, where they could be plucked up by large birds of prey. At the sound of the 'snake' alarm, all the members of the troop stand on their hind legs and search the grass for the intruder.

The calls and the correct responses must be learned and young monkeys do not always get them right. One youngster, prompted by a falling leaf, was observed to give the eagle alarm, but realised his mistake when none of the adults in the troop ran for the safety of the bushes.

Colobus monkeys also warn each other of approaching danger, but to give a loud call might, in certain circumstances, make the situation worse. This species of African monkey has two main natural predators, apart from man: the leopard and the chimpanzee. In the Thai Forest of the Ivory Coast, red colobus monkeys have to contend regularly with both predators, but have a different defence strategy for each. In the case of a leopard, the monkeys scream their warning calls – which can also be heard by the predator – whereas a troop

of hunting chimpanzees is greeted with silence as the chimpanzees could chase the monkeys by following their cries.

Another safety strategy for the colobus monkeys is to team up in the forest with diana monkeys. Diana monkeys are too agile for chimps to catch, but are vulnerable to predation from crowned hawk eagles, which snatch them from the canopy. They are therefore highly alert and popular sentinels in the forest (several species of forest monkey team up with them for this purpose). Colobus monkeys are leaf-eaters and do not compete for food with the mainly fruit and insect-eating diana monkeys. The dianas become the eyes and ears of the red colobus and provide early warning of approaching groups of hunting chimpanzees.

SAFETY IN NUMBERS

Many pairs of alert eyes in a flock, herd or shoal is a dependable survival strategy, but how do the individuals in a flock of 100 birds or a shoal of 1000 fish escaping from a predator avoid chaos and confusion as they swirl backwards and forwards? The secret lies in a chorus line and a flash of shiny scales.

A flock of flying birds behaves like a high-kicking line of dancers. When a predator, such as a fast-flying peregrine, swoops on a

flock, and a bird banks into the flock to avoid attack, its immediate neighbours turn with it. The change of direction is registered by one bird, then another, with increasing rapidity until all the birds have completed the manoeuvre. The birds nearest the initiator respond slowly because they rely only on their natural reaction times. Birds further away see the approaching wave of movement and time their response to coincide with its arrival at their own position. By anticipating the manoeuvre, they can change direction at up to three times the speed of their normal reaction time. In a flock of dunlin, a member of the sandpiper group, birds close to the initiator take about 67 milliseconds to respond, while those at the end of the line take only 15 milliseconds.

A manoeuvre takes place only when a predator moves into the flock. If the predator attacks from the right, the birds on that side turn left towards their neighbours, bringing about a change of direction in the whole flock. If a bird turns away from the flock, it is ignored. This difference in reaction, it is thought, is a simple response to predation. Birds of prey tend to pick off individuals that have become separated from the flock. To avoid being left out in the open and vulnerable to attack, birds will only follow an initiator that moves into the body of the flock; indecision is potentially fatal.

The large-scale movements of the flock also deter predators. A bird of prey takes care not to get caught up with the flock it is attacking as a midair collision could be disastrous. A tightly wheeling formation can therefore avert an attack: the tighter the flock, the safer each bird will be, and there is an instinctive drive for the centre. Sometimes, however, the chorus line splits instead. Each section flies to one side of the predator, and the predator has to concentrate on avoiding a collision.

The same rules apply to a shoal of fish, with the addition of mirrors. The silvery surfaces of fish — consisting of about 1 million reflecting platelets per square $^1/_2$ in (1 cm) – reflect the sea around them, making them effectively invisible, but when a fish rolls to one side its scales reflect the light from the sea's surface. If it rolls to the other side the darker depths are reflected.

SYNCHRONISED SWIMMING   To keep together, fish in shoals observe the flashing scales of other fish and listen for the sounds of sudden movements.

The platelets, however, are angled so that they catch the light more efficiently, and one fish swimming beside another will see changes in brightness as its neighbour turns, rolls, pitches or yaws. In order not to collide, neighbours react by moving in the same way, with a reaction time 20 times faster than that of humans. The effect, much like the bird-flock chorus line, travels through the shoal, which moves as one.

Fish such as mackerel and horse mackerel also have special plates on the tail and other parts, which reflect differently coloured light. As they speed up or slow down, the plates signal the change to the rest of the shoal. When a predator approaches, the sudden changes of colour as the shoal manoeuvres cause it maximum confusion.

There is speculation, too, that low-frequency sound plays an important role in group cohesion. When whiting and herring change direction, the movement is preceded by brief pulses of sound produced by the swim bladder. The sound does not travel far, but neighbouring fish in a shoal should be able to detect it using the vibration-sensitive organ in the head and along the side of the body, known as the lateral-line system. These sound pulses alert neighbours that a movement is imminent, and so fish in a group are primed to move quickly and in unison.

Flocking or shoaling with animals of the same kind may be one strategy for survival, but more vulnerable species sometimes team up with more aggressive or larger species, the latter affording some degree of protection to the former. Weaver birds choose nest sites close to wasps' nests. Gulls sometimes nest close to ever-alert terns. And small groups of Thompson's gazelles, the ubiquitous prey animal of the savannah, join up with herds of the larger Grant's gazelles: the more numerous the herd, the greater the protection for the individual.

THE FIT SURVIVE

Gazelles and antelopes have another trick to confuse a lion, leopard or hyena – they pronk. If danger is close by, they will leap vertically into the air and bounce around with stiff legs. It is demonstrating that it is too fit to catch.

Even relatively slow creatures such as butterflies use fitness and manoeuvrability to escape predators, but they have also come up with a variety of protective adaptations. Some butterflies make themselves unpalatable to predators, and especially to birds, by producing noxious, and sometimes toxic, chemicals within their bodies. Others simply out-fly their attackers. In the palatable

TOO FIT TO EAT *By showing
its ability to leap, an impala
signals that it is not easy to
catch and that a predator
should look elsewhere.*

butterflies, such as morpho butterflies, up to 40 per cent of the body mass consists of flight muscles, whereas in their unpalatable relatives it is as low as 22 per cent. As a consequence, the great majority of edible butterflies and day-flying moths can accelerate faster than birds, while some of the more poisonous species can barely lift off the ground. It is all the more remarkable that the palatable butterflies, like their warning-coloured inedible relatives, sport bright colours. The morpho butterflies, for example, have conspicuous iridescent blue and black upper surfaces to their wings. It seems that bright colours in highly manoeuvrable species that mimic each other are a way of notifying predators that it is not worth trying to pursue them.

CHEMICAL DEFENCES

Another common survival strategy in the animal world is to threaten a predator with something nasty. The skunk has brought this to perfection. Its infamous spray, according to the English naturalist W.H. Hudson, 'pervades the whole system like a pestilent ether, nauseating one until seasickness seems almost a pleasant sensation by comparison'. Hudson was in no doubt about the blinding and choking defensive

CHEMICAL ARSENAL
*A pericopid moth (left)
discourages predators with an
unpleasant foam of foul-
smelling chemicals that will
evaporate into a repugnant
mist. Right: The sleek sheen of
morpho butterflies tells
predators that they are fit,
fast and not worth chasing.*

discharge of the world's smelliest carnivore. On one of his many journeys in Latin America, Hudson was on the receiving end of the sticky yellow spray – the skunk's ultimate weapon. But he may well have taunted this normally docile mustelid (a relative of badgers and otters), since it takes a good deal of provocation for a skunk to live up to its disagreeable reputation.

Skunks are active mainly at night, searching for insects and the small animals on which they feed. They prefer to forage along the edges of forests and move in a leisurely, almost nonchalant way with little apparent concern for danger. The conspicuous black-and-white pattern of the fur serves, like the black-and-yellow stripes of the wasp, as a warning and is the first line of defence to deter predators. The various species have different patterns: hooded and hog-nosed skunks have a black body and white back, striped skunks have forked white stripes along the body, and spotted skunks have broken white stripes and spots.

If this colour code fails to impress a predator and the skunk is actually threatened, it will bring its second line of defence into action, reinforcing the warning. The large striped skunk arches its back, stamps its front feet on the ground, shakes its head from side to side, raises its long-haired tail vertically and struts around stiff-legged. The smaller, more agile spotted skunk has a more exaggerated display – it performs handstands

on its front feet and walks a few steps with its hind legs and tail held high in the air.

And if that does not work, the skunk returns to all fours, bends its body in a U-shape, turning its rump towards the predator, lifts its tail and squirts. Highly volatile, sulphurous compounds are stored in a pair of enlarged anal glands surrounded by muscles that can contract at will. The musk, common to many mustelid animals and more usually used for attracting a mate, is forcibly ejected through two tiny nipples inside the anus. The fine spray can hit a target with considerable accuracy at 6 ft (1.8 m) but can reach to 21 ft (6.4 m) with a following wind.

The secret ingredients of the musk itself has only recently been analysed. It consists of a noxious cocktail of seven evil-smelling liquids. The ingredients are thought to attach chemically to the proteins in hair and wool, making them very difficult to remove. One of the chemicals has the property of reacting slowly with water. Therefore, a victim's sweat can help to release more of the odour for days. The skunk aims at the face, the noxious fumes

choking a predator and sometimes causing temporary blindness. The smell can wrinkle a nose up to $1^1/_2$ miles (2.5 km) downwind, but not all predators are discouraged. Eagles and owls, particularly the ferocious great horned owl, take skunks; so do pumas, coyotes, badgers, lynxes and foxes, but they have to be extremely hungry before they risk a spraying.

Many creatures, including some of the lowly insects – particularly beetles – produce noxious chemicals for defence. The star performer in this entomological theatre of war is the bombardier beetle, *Brachinus.* It more usually hides under stones, but if

caught in the open it is able to hit predators approaching from any angle with a boiling blast of noxious chemicals and live to fight another day.

If an ant grabs it by the leg, the bombardier beetle is able to swivel its abdomen around in a split second and direct the jet of scalding liquid straight at the attacker. Under continuous attack, the beetle is able to fire 20 or 30 times, each discharge accompanied by an audible report and a cloud of vapour visible from a pair of openings at the tip of its abdomen. The boiling hot spray is the result of a chemical reaction in a tiny chamber in the beetle's rear end. Hydroquinones are mixed with hydrogen peroxide and a cocktail of enzymes to produce oxygen, which acts as a propellant, and a chemical irritant that is released at 100°C (212°F), the boiling point of water. Researchers studying the beetle and its

BAT WARS

Moths have ears with which they are able to detect the sonar beams of hunting bats. They may be on the thorax, abdomen or head, or, in the case of hawk moths, on the mouth-parts. Ears most likely evolved to monitor wing-beat vibrations, giving the insect an in-flight checking system.

The ears are very simple, with just two nerve cells (compared with 17 000 in the human ear), but they are effective in detecting an incoming predator. A faint bat is a distant bat and a loud one is too close for comfort. With a tympanic organ (an auditory membrane, equivalent to the eardrum in humans) on each side of the body the moth can also detect whether the bat is coming from the left or the right.

Having detected an attacker, the moth then attempts to shake it off in a series of aerial manoeuvres. It flies in large loops and circles, but if the bat swoops to within 3 ft (1 m)

the moth deploys its secret weapon, producing several loud bursts of a high-frequency, rasping sound by activating tiny plates on either side of the body. The sound, with an average frequency of about 20 kHz, startles the bat.

The sound emitted by the moth is similar to the bat's returning echo, and the bat, which at that moment is probably diving at maximum speed, banks away to avoid the phantom 'obstacle' it has suddenly detected. In the confusion the moth escapes.

Moths, however, are not having everything their own way. In this evolutionary arms race, the bats are beginning to adopt counter-measures. Several species which normally chase night-flying moths have raised or lowered the frequencies of their echolocation signals in order to be outside the sound spectrum detected by moths. Moths are at present sensitive to

sounds between 25 and 50 kHz, and so a bat emitting sonar signals above 60 kHz or below 15 kHz is able to get to within 6 ft (1.8 m) of the moth without being detected.

Other night-flying insects, such as green lacewings and praying mantises, are also vulnerable to bat predation, and they have ears too. The delicate green lacewing has an ear on each wing, while some, but not all, praying mantises have a single ear underneath the body.

The ear is most sensitive to the frequencies of sound emitted by hunting bats. Having spotted the pursuing bat, the praying mantis also launches into an aerobatic display in order to outmanoeuvre it. It flies suddenly upwards, stalls, and drops downwards in a power dive. The bat flies after it, the two of them hurtling towards the ground. Eventually, the bat pulls out of the dive, while the mantis lands gently and waits for the danger to pass.

chemical defence system also discovered that the chemical discharge is not continuous but pulsed. Instead of coming out as an even stream, the spray squirts like a machine gun at a rate of 500-1000 pulses per second. The explosion chamber cools momentarily between pulses so that the beetle does not blow off its own behind.

The chemical defence system of the paussine bombardier, *Goniotropis*, is less sophisticated but still effective. The openings for the chemical spray are spaced out along the side of its body, near the tip of the abdomen, but instead of waggling the abdomen the beetle uses a principle of physics known as the Coanda effect. When milk is poured from a jug it sometimes has the annoying habit of curling round the lip and dribbling onto the tablecloth. The principle – that gases and liquids tend to follow the curvature of a solid – was named after a Rumanian engineer, Henri Coanda. In the case of the beetle there is a flange on either side of the abdomen. Even if the insect is facing a frontal attack, it can bend the liquid through a full 50 degrees to meet the attacker and it can swivel its abdomen slightly to deal with attackers at the side and rear.

REAR SPRAY *The Eleodes beetle stands on its head to use its chemical weapon, spraying a concoction of noxious substances at an adversary.*

A metriine beetle has an even simpler system. It ejects boiling, poisonous fluids from beneath its wing-cases. During an attack the beetle froths in a simmering, bubbling mass of noxious chemicals that sends droplets in all directions. A cloak of vapour enables it to walk nonchalantly through a swarm of would-be predators, such as ants. Not all predators are kept at bay by chemical defences, however. The *Eleodes* beetle of the Arizona Desert deploys its artillery by standing on its head and squirting a noxious mixture of cold quinones at the attacker. However, the grasshopper mouse deals with this by grabbing the beetle and pushing its rear end into the sand, while it feasts safely on the front.

THE BORROWERS

Unable to produce and deploy their own chemical defence systems, some animals borrow noxious substances from plants and other animals. Many plants produce defence chemicals, such as alkaloids or phenols, and certain insects appropriate them for their own defence.

The larvae (caterpillars) of Australian sawflies use eucalyptus oils. They hatch from their eggs and are guarded by their mother until they have taken in enough oils from their food to fend for themselves. Nutrients from the leaves pass to the stomach, but the resins are somehow separated and diverted to pouches in the foregut and stored. During the day, when they are more vulnerable to birds, the larvae adopt a rosette formation, heads facing outwards, and, if threatened, regurgitate the oils. If an individual is separated and pursued, it will bend its head back and daub oil onto its back as protection. The smell is strong enough to put off all but the most zealous predator.

The camphor assassin bug gathers the resin of camphor weeds. Very few insects can tolerate the minute droplets of this sticky substance, which forms on the plant's leaves and stalks, but the assassin bug protects its eggs with the resin. The female collects the resin on her front legs, wipes the secretion onto the middle legs, and then, using the hind legs, plasters it over her underside. As a consequence, every egg she deposits is liberally coated with camphor to protect it from egg predators.

Bitter-tasting heart poisons, known as cardenolides, are present in the sap of milkweed, and these effectively protect the

caterpillar, chrysalis and adult monarch butterfly from birds. The poisons are acquired from milkweed plants by the caterpillar and retained by the adult insect. The monarch butterfly warns potential predators of its chemical defence by wearing bright orange-and-black colours. Insect-eating birds, having tasted one butterfly, are reluctant to try another. Monarch butterflies vary in their unpalatableness, some containing large amounts of cardenolides and others little. Only one in three monarchs is unpalatable. Their gaudy colours warn away most passing birds but two species, the black-backed orioles and the black-headed grosbeaks, which frequent the butterfly's winter home in Mexico,

CHEMICALLY DEFENDED *The bright colours of the monarch butterfly warn insect-eating birds that many are foul-tasting.*

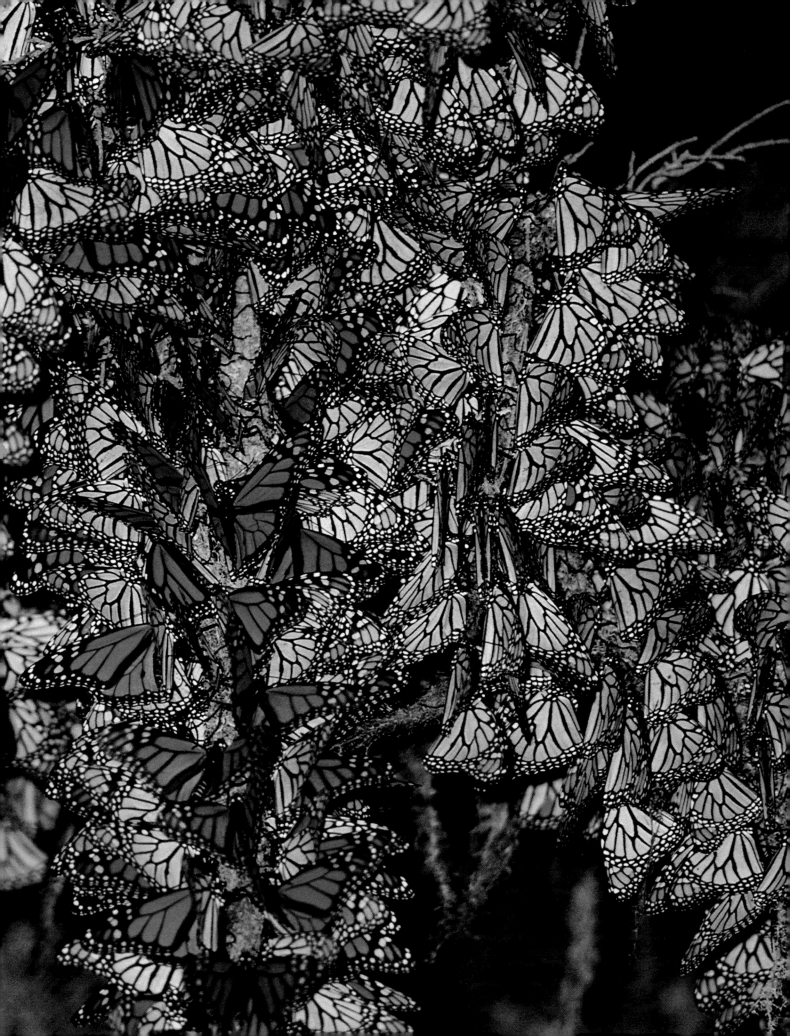

SAWFLY ROSETTE *To deter predators, sawfly larvae regurgitate substances, such as pine resin or eucalyptus oil, acquired from trees.*

and openings in the thorax. Air from the respiratory system reacts with defence chemicals, creating the foam. Thousands of tiny bubbles in the foam then burst, releasing a fine cloud of repellent around the creature's body. If the predator is not deterred, the grasshopper's second line of defence is to vomit a particularly smelly

slurry. Most attackers make off rapidly. In attempting to analyse the chemical content of the grasshopper's defence system, scientists discovered a chlorinated phenol, a substance more usually associated with pesticides, in the regurgitated froth. The grasshopper, it is thought, had eaten the man-made chemical and incorporated it into its own defences. The secondary use of an artificial pesticide is an example of nature's nasty habit of picking up abnormalities in the environment and turning them to its advantage – but not necessarily to the benefit of humans. The episode illustrates the potential dangers of using chemicals in the environment.

ignore the warning. The orioles taste the butterflies, rejecting the poisonous ones and swallowing the others, while the grosbeaks simply eat the lot, poisons and all.

For those animals incapable of producing or storing poisons or nasty chemicals, the next best thing is to look like one that can. Viceroy butterflies, without poison, mimic poisonous monarch butterflies, and in doing so acquire some degree of protection. Hoverflies mimic the yellow-and-black stripes of wasps and bees, and the yellow-and-black death's head hawk moth resembles a large bee. The hornet clearwing moth looks, to all the world, like a hornet. Its wings even lack the normal covering of scales on a moth's wings and are translucent, and its body has the same striped warning pattern as the hornet's. Similarly, North American king snakes have patterns of red, black and yellow rings nearly identical to those of the venomous coral snakes. The only difference between the two species is in the order of the stripes.

One story of chemical defence has far-reaching implications, not for insects but for people. In the southern United States there is one grasshopper that does not behave like others. Most grasshoppers are fast moving, but the lubber grasshopper is large, flightless and slow. Under attack it does not use its long hind legs to jump and escape, but hisses and froths at the mouth

LAND LUBBER *The lubber grasshopper defends itself from predators by producing a noxious foam from its mouth and respiratory spiracles.*

CONTINUING THE LINE

5

CHOOSE ME! *The plumage of the cock great argus pheasant proclaims his fitness to mate.*

THE SOLE PURPOSE OF AN ANIMAL'S EXISTENCE FROM BIRTH, THROUGH HATCHING, GROWING UP, LEARNING, MOVING, EATING, COURTING AND MATING, IS TO PERPETUATE ITS OWN GENES. HOWEVER, INDIVIDUALS ARE FAR FROM BEING CONCERNED ABOUT THE FATE OF THEIR OWN SPECIES; INSTEAD, THEY ARE UNEQUIVOCALLY SELFISH. MALE ANIMALS, AND OCCASIONALLY FEMALES, WILL FIGHT RIVALS FOR THE RIGHT TO MATE. THEY WILL ALSO FIGHT FOR ACCESS TO A TERRITORY IN WHICH VITAL RESOURCES, SUCH AS FOOD AND SHELTER, ARE PRESENT, ENABLING OFFSPRING TO DEVELOP AND GROW.

MATING BALL *Male red-sided garter snakes vie for a female.*

THEY WILL CREATE AN ENVIRONMENT, WHETHER IN AN EGG OR A WOMB, IN WHICH THEIR OFFSPRING CAN OBTAIN THE BEST POSSIBLE START IN LIFE.

RIVALS AND ALLIES

Life is a constant contest – for living space, food and a mate.

Many animals attempt to exclude others of their own species as they are direct competitors for the same ecological niche. Some animals, though, enlist the help of friends.

An animal needs somewhere to live, a safe place with sufficient food to attract a mate and raise a family. The best sites are usually at a premium, and so, through fights and feuds, the strongest and fittest gain the best territories. For some animals, such as the tiger, leopard and some prides of lions, a territory is permanent, defended and retained throughout an animal's life. For others, such as songbirds, a territory lasts for a summer only. Each year a male must establish his rights, usually by singing morning and evening, to a particular plot.

CLAIMING A TERRITORY

The poets would have us believe that birds singing at dawn are celebrating the start of a new day. Nothing could be further from the truth. In reality the dawn chorus is a daily outbreak of war. At dawn the sun, and air temperature with it, has still to rise; for a songbird this is not the ideal time to forage. The availability of prey – mainly insects – is generally low, and so a bird has time to spare. Territory holders tend to die during the night and there are vacancies by morning. A wanderer would do well to be on

DAWN SOLO *At dawn a cock blackbird declares to others of its kind that it is in residence in its territory and woe betide any intruders.*

the wing at this time. A resident must therefore be alert to intruders, for it is at dawn that they are likely to be most active. The resident's 'keep out' sign is his song, which he sings over and over again.

Dawn is a better time of day than dusk to sing out a proclamation of territorial rights, for the air is often still and conditions are generally favourable for maximum sound transmission. The quality of the song is important, for the notes contain clues about the well-being of the singer. Male songbirds with territories rich in food tend to sing louder and for longer, indicating their strength and fitness – factors that help not only to discourage rivals but also to attract a partner. Indeed, the quality of a song is dependent on how well a bird has fed the previous day.

The behaviour of a songbird's mate will determine how and when he sings. The male blackbird sings with gusto to repel rivals and keep them at a distance from the female, but when she leaves the nest he stops singing and follows her closely. They forage for the best part of the day, but he

MALE POSTURING
The cock great tit (above) is often up before dawn, declaring his territorial rights and singing to his partner. Right: The bull hippopotamus opens his mouth wide in order to impress on other hippos that he is king of the river.

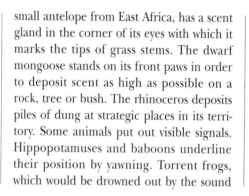

returns home ahead of her and starts his evening song. When she is safely on the nest he stops singing again. As she reaches peak fertility – the time of greatest danger from rival males – the cock blackbird sings with even more verve. And each day he sings progressively earlier until she lays her eggs, and then he stops singing.

Male great tits are active earlier than blackbirds. The female lays her eggs at dawn, but is receptive to mating again about an hour after dawn. The cock bird must therefore ensure that he is the only great tit male around, and so he sings just before, during and after dawn, not only to discourage rivals from entering his territory but also to woo his partner.

Smell, too, is an important territorial marker. Hyenas, wolves and buck antelopes scent-mark territory with urine and secretions from special glands located on the rump, the feet or the face. The oribi, a

small antelope from East Africa, has a scent gland in the corner of its eyes with which it marks the tips of grass stems. The dwarf mongoose stands on its front paws in order to deposit scent as high as possible on a rock, tree or bush. The rhinoceros deposits piles of dung at strategic places in its territory. Some animals put out visible signals. Hippopotamuses and baboons underline their position by yawning. Torrent frogs, which would be drowned out by the sound

HEAD-TO-HEAD *Bighorn sheep behave like battering rams during the rut, but have shock absorbers in the skull to prevent brain damage.*

of a nearby waterfall, have lost their croak and instead shake a bright blue foot. Fiddler crabs wave a single giant pincer in the air. But whatever the means of communication, the message is the same: keep out!

THE FIGHT

Rival males often fight for the right to mate, but they will first try all manner of means to avoid a brawl. Ritualised combat may prevent rivals from being seriously damaged, but if the ritual fails a real fight follows. Red deer, for instance, roar first and battle later, but they will only do so if the contestants can match each other roar for roar. The rate of roaring indicates the fitness of an individual, and only stags roaring at the same rate lock antlers in push-and-shove trials of strength.

Several other animals have

HEADBANGERS *European male ibex crash headlong into each other, fighting for the right to mate with the local females.*

NECK-TO-NECK *A pair of young male Rothschild's giraffes engage in a ritualised fight by banging their necks together.*

developed special weapons for combat, often on the head or face. Male klipspringers have short, sharp horns and jab at each other; reedbucks have horns with forward-curving tips, with which they can gently push each other to the ground without slipping; red deer stags have many-branching antlers which they can interlock with those of their rivals.

The most impressive displays of rivalry must be those of the 'headbangers'. European ibex butt each other horn to horn, like other mountain sheep. North American bighorn rams rise on their hind feet, topple forwards and slam their heads together – the sound of the impact reverberates around the crags and valleys of their Rocky Mountain home.

North American bison avoid a fight if at all possible. Opponents threaten and yield with signals that are intended to encourage a rival to back down before a real and perhaps damaging altercation. Rivals may bellow at each other, urinate in the dust and roll in it, or approach each other head on. They may stand close together in a non-threatening way and nod their heads in unison. One may back down by withdrawing or nonchalantly munching some grass. If neither bull retires, a fight ensues. The two rivals clash heads, the horns of one sometimes ripping out handfuls of hair from the other. They circle, each trying to out-dance the other, or to hit the vulnerable flank of his opponent. A violent hit, broadside into the rib cage, could mean death for the weaker of the two combatants.

Nevertheless, violence at breeding time is prevalent among male animals. Elephant bulls stab and push at each other, using huge tusks. A wrong move can result in a punctured throat or lung, and certain death. Male narwhals – the unicorns of the sea – fence with an elongated tooth, kangaroos box with their powerful hind legs and giraffes crash their extraordinarily long

CHANGING SEX

Some animals have the best of both gender worlds: they change sex. For animals that live in isolated groups, such as clown fish, this ability to juggle the sexes could mean the difference between survival and extermination, for if the single male of the group were killed it might be difficult for another to take his place. This ability to change sex is not uncommon in the sea.

The Atlantic silverside is a fish that is common in estuaries along the Atlantic coast of North America, but populations from South Carolina differ from those farther north in that their sex is determined by the temperature of the water at the moment they hatch. In warm water, about 21°C (70°F), the silverside is male, while in cold water 15°C (59°F) it is female. The curious adaptation apparently has survival value.

Along the South Carolina coast, where the growing season is long, in the cold waters of early spring female fish are produced. They have spring and early summer to feed well and get themselves into good shape to survive the following winter. In addition, the larger the female grows, the more eggs she can produce. A small male, however, can produce sufficient sperm to fertilise all his partner's eggs, and so males are produced later in the growing season when the water is warmer.

On the European side of the Atlantic, members of the wrasse family go one better: they can be born one sex and later change to the other. The ballan wrasse, for example, relies totally on sex change to produce males of the species. In the first 14 years of an individual's life, it is a female, and only in 'middle age' does it change to a male. Not all females make the transformation,

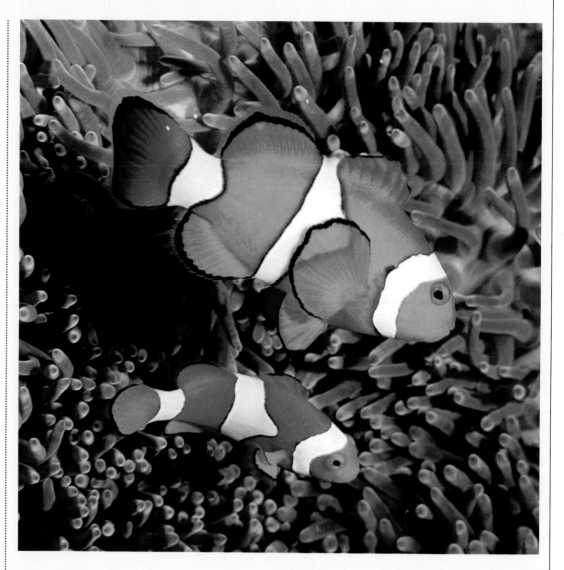

however, for there is only one male to every ten females.

On Australia's Great Barrier Reef, the pygmy angelfish (Centropyge bicolour) lives in a harem consisting of one male and between five and ten females. If the male dies, the largest female changes sex and takes over the harem. When the number of females exceeds ten, the largest becomes a male, takes some of the females and starts a new harem in another territory.

Male clown fish – small, colourful

fish that live among the tentacles of sea anemones – can change the other way, from male to female. A group consists typically of a large female, a smaller male and several juveniles. If the female is killed or dies, the male changes sex and one of the juveniles becomes a fully functioning mature male.

Sex changes occur in species of fish in which individuals of one sex are particularly successful in reproduction. A large male, for instance, would acquire, retain and mate with the

SEX TRANSFORMATION *The smaller male clown fish can change into a large female if his mate dies or is killed by a predatory reef fish.*

breeding females, to the exclusion of smaller males. In this situation it makes good sense for a fish to be female when it is small, yet still in demand, and male when it grows bigger and can compete with the other large males. Sex change allows a fish to get the best of both worlds.

necks together. Bull elephant seals attack each other with extraordinary ferocity. Sharp-toothed jaws rip into blood-streaked bodies, as they crash heads and necks. An area of thick blubber around the neck and upper part of the body protects vital organs from serious damage. The male rhinoceros, too, has specially thickened skin on its flanks and back to ward off blows from a rival's horn.

THE ARENA

Formal competitions intended more to attract the attention of the females of the species than to intimidate rivals may take place in traditional display arenas or leks. Males compete for the best places at the centre of the lek, where the females congregate, as males occupying these prime sites are more likely to attract partners than those at the edge. And when the time comes to perform, a young beau gives his all.

Prairie chickens dance, boom and stamp their feet, capercaillie strut and gurgle like liquid emptying from a bottle, and humpback whale bulls sing the longest songs known to science and dance beneath the sea. Male dugongs, a species of sea cow, space out and hang vertically in the sea. And Ugandan kobs, a species of antelope, create a patchwork of smelly sites, the smelliest patches being the most attractive to the hinds. The smell, however, is not produced by the males at the lek but by the females themselves.

Male kobs take up and defend small territories, each about 65 ft (20 m) across, in the lek, which is made up of 30 to 40 territories. During the breeding season, males come and go, and several may use the same display site. The females, however, seem to be attracted not to the male in the site but to the site itself. The same few sites attract the most females. It seems that a component – still unidentified – in female urine attracts other females to the site. Female kobs, it appears, are very conservative about selecting a mate, preferring to choose sites

DISPUTE ON THE DANCE FLOOR
Cock prairie chickens from North America fight for prime positions in the dancing arena or lek.

PASSIVE DISPLAY *A male dugong visits a lek to impress females, but forgoes a performance, preferring just to hang in the water.*

that have been used many times before. These sites, proven safe from predators, are also fought over by males, the fittest winning residency, albeit temporarily.

Travellers on certain West African rivers may be greeted by the extraordinary sight of hammerheaded bats performing at their lek. All along the river bank, for about 1 mile (1.6 km), rows of bats hang in the trees, making a horrendous noise and frantically flapping their wings in the hope that their bizarre displays will attract a passing female – the noisiest and most vigorous males attracting the most attention.

The process starts when the males leave their daytime roosting sites, fly to the riverside lek and, early in the wet seasons, from June to August and again from January to early March, jostle and fight for prime positions in the line. They are at the lek every evening, hanging from the trees in two formal rows, one on each side of the river. At first there may only be a few animals present, but at the height of the season there may be as many as 150 fruit bats, each separated from its neighbour by a 40 ft (12 m) gap. The male bat, with his huge hammer-shaped head, is twice the size of the foxy-faced female. In order to make his loud calls, the male has several anatomical adaptations that are absent in the female. His chest cavity is filled by a large, bony larynx or voice box, the cheeks are pouched, the nasal cavities can be inflated, the mouth is funnel-shaped, and there is a grotesque nose-leaf on his face. The metallic-sounding call is very loud and is given between one and four times a second.

When a female arrives at the lek she flies up and down the river, like a shopper at a bazaar. She hovers in front of males that take her fancy, but often cannot decide whether to buy or not. Any male under close scrutiny pulls out all the stops and gives a loud staccato buzzing call. If the female is impressed, she does not let on at first. Instead, she flies up and down, hovering tantalisingly in front of this male and that until she returns to the same few males time and time again. Eventually she selects a partner and perches upside down alongside him. Mating is completed after no more than 30 seconds, and the female flies away. Often it is the same few well-placed and noisiest males who attract all the females.

FORMING ALLIANCES

Male animals do not spend all their time in competition with others. Youngsters may find it useful to bide their time and learn from their elders. Relatives may get together

RIVERDANCE *The male hammerhead bat hangs about in trees along West African rivers and tries to display more noisily than his neighbours.*

BOWER THIEVES

Male bowerbirds build to impress females. Some simply clear a court, others put down a mat, and still others erect elaborate maypoles or construct great avenues. The tooth-billed catbird (*Scenopoeetes centirostris*), for example, clears a stage on the forest floor and decorates it with leaves, carefully placed with their lighter undersurfaces uppermost. Archbold's bowerbird puts down a soft mat of mosses and ferns.

A simple maypole bower, such as the one MacGregor's bowerbird builds, might consist of a sapling surrounded by a conical-shaped column of interlaced leaves, with a platform of mosses and bearded lichens at its base. The bower too is decorated with lichens – as well as charcoal, fungi, seeds and parts of insects. The column is often draped with insect frass – the silk-like material mixed with sawdust which is produced by the woodboring larvae of moths – and a border of screw pine needles surrounds the base. If a decoration should fall, the bird immediately replaces it in exactly the same place. No two bowers are alike, and they may be up to 10 ft (3 m) tall. Each season the bird adds a little more to its bower.

Avenue bowers consist of two parallel walls of twigs and grass stems set on a platform of sticks.

The walls are painted with charcoal moistened with saliva – this is applied with a piece of bark. The walls built by the satin bowerbird curve in at the top, forming an arch that frames a large collection of objects, all of them blue or green – matching the sheen of the plumage. In recent years, bowerbirds have been supplementing their natural decorations, such as parrot feathers, with car keys, bottletops, toothbrushes, shotgun cartridges, spectacles and just about anything of the right colour that catches the bird's eye. Some birds have even been known to paint their bowers with blue washing-powder granules. The bower is not a nesting site but simply a place to which a male can entice a female to impress her, and competition for the most extravagant bowers is fierce. At every opportunity, a male satin bowerbird will attempt to destroy his neighbour's bower or steal his decorations.

The satin bowerbird orientates its bower with the long axis 30 degrees, at the most, off the north-south line. This ensures that neither sex has the sun in its eyes during the frantic morning courtship performance. During the display the male tempts the female into his bower by picking up objects, gyrating, bowing and scraping. She is understandably

nervous, for mating is a rather rough affair during which the entire bower can be demolished.

The female eventually crawls away bruised, battered and exhausted, and builds her nest well away from the bower site. She brings up her brood on her own, while the male rebuilds his bower and begins the ritual all over again with a new mate.

BOWER-BUILDER *The cock MacGregor's bowerbird constructs a palace of love with twigs and leaves to attract a female.*

and collaborate in order to perpetuate their line. Some males forsake the battle and, like the bottle-nosed dolphin, put aside their differences to become allies. Dolphins normally form into loose groups, with animals frequently moving between groups as they break up and reform. But the most stable groups consist of two to three males, and these individuals stay together for a long time. Their unexpected camaraderie is most apparent when a receptive female is close by. She may be suspicious of her three

suitors, who might have to chase her for miles. When they catch her, they perform a series of elaborate underwater displays that would be the envy of any synchronised swimmer. They move in formation, twisting and turning in unison, and leaping into the air. Having impressed and 'captured' their female, the three escorts keep her secure with a formation that boxes her in. They might swim in this formation for up to four weeks, during which time the males attempt to mate with her.

If the courtship group is confronted by another group of males, the outcome is not predictable. The interlopers may be intimidated by the residents and leave quietly. Or they might team up with another bachelor group and chase off the residents, thereby acquiring the female for themselves. The alliances, however, are not permanent. Group A might form an alliance with group B and chase off group C one week, and then a few weeks later join with group C and see off group B.

COURTSHIP AND MATING

Animals belonging to the same species have to compete with each other for living space. In order to mate and to raise a family, however, they must suppress their natural urges to fight or flee, and in this courtship rituals hold the key.

Once they have established a territory, the males have to woo the females, not for any romantic reason, but to overcome the natural tendency of an animal of the same species to see another, no matter what its gender, as a competitor. He must also impress a female into believing that he is the fittest partner available and will contribute the best genes. Females, then, tend to have the right to choose, and it is up to the male to impress.

The more eyes a peacock has on his tail, the more likely he is to impress females, although the peacock's tail may be an example of evolution carried to extremes. However, many animals, particularly birds, use extravagant displays to demonstrate their prowess as partners and their fitness to sire the healthiest offspring. Argus pheasants have tails that rival those of peacocks, and birds of paradise match their lavish plumage with even more extraordinary dances and displays. For other birds, such as the African whydah, the secret lies in the length of the tail.

Male whydah or widow birds guard territories well stocked with food and nesting bushes. The fittest birds have the best sites, but females would have their work cut out to find them because the territories are so widely spread out. The males assist them by sporting tail feathers that inform the hens at a glance which is top cock.

SEDUCTIVE EYES *The peacock's tail persuades a peahen that its owner is fit and healthy, and able to survive predation despite the inconvenience.*

A male whydah is jet black, with broad tail feathers that are several times the bird's length. Females fly over the grasslands searching for the local talent, and when a cock bird spots one he jumps about 3 ft (1 m) into the air from a specially cleared display arena. Fluttering wings and streaming black tail feathers attract the attention of the passing hens, who are looking for birds with long feathers, which indicate a long life. A minimum of tearing and parasite damage in tail feathers is also a sign of the bird's good health, so cocks with the most impressive tails get to mate with the hens.

The reason for the male's extravagant and often flamboyant displays is that he is usually competing against several males for fewer females. The females do the choosing, and a peacock or a bird of paradise that can avoid being caught by predators, encumbered though it is by awkward adornments and predator-friendly colours, must be a survivor.

Another theory holds that a flamboyant male is trying to impress not only females but predators too. The bright plumage and long display feathers of a cock bird might work in the same way that the antelope's 'pronking' tells a predator that it is very fit and might well get away if chased. The tail of a peacock is an obvious burden in flight but may serve to suggest

ALLURING APPENDAGE
Tail length for the male long-tailed widow bird or whydah (left) is critical in attracting a mate; the longer the tail, the fitter the bird. Right: The male pintail whydah seduces a female by showing off his tail during a ritualised display.

pond so that his voice sounds deeper. To a female toad a deeper voice means a bigger, stronger mate.

Sound is particularly important to courting crickets and grasshoppers, the former stridulating, producing its shrill chirping sounds, by rubbing its forewings together and the latter by scraping its hind legs against its wings. Some species go to extraordinary lengths to make their call louder: the mole cricket builds a tunnel the shape of a megaphone to enhance its call – it is audible to the human ear at a distance of over 1 mile (1.6 km) – and the tree cricket nibbles a hole in a leaf, which acts like a baffle to amplify its call.

The tiny male fireflea of the Caribbean – 1/16 in (2 mm) in length – achieves his sexual ambitions with a flash in the dark. As he swims about under water, he squirts out a cocktail of chemicals from glands on his upper 'lip'. This combines with the seawater to release energy in the form of a bright blue light. In fact, the light is so bright that a crushed fireflea provides enough light for a person to read a newspaper in the dark for up to ten minutes. The male fireflea attracts a mate by leaving a trail of glowing blue dots behind him. Each dot is no bigger than the fireflea himself, yet it can be seen under the water for 30 ft (9 m) or more and glows for about 15 seconds, during which time the fireflea hopes to catch the eye of the one female in a hundred that is fertile and ready to mate. Sneaky males sometimes hide in the glowing

wake of others in the hope of winning a mate without going to the trouble of producing a light of their own.

THE MATING GAME

Life's dangers do not end with finding and attracting a mate. Having been chosen or having done the choosing, the two sexes must come together to mate. Some partners are less than friendly – even outwardly hostile. The female praying mantis, for example,
continued on page 144

LOST HIS HEAD *The body of the male praying mantis (below) continues to mate with the female even after she has eaten his head. Overleaf: A pair of young albatrosses perform their courtship display for the first time at five years old.*

that the bird is fit enough to survive, even so. And if a predator gives chase but fails to catch a distinctively marked bird, it might remember not to waste energy on it in future.

Whatever the reason, male animals employ all manner of means to make themselves stand out. There are callers such as frogs and toads, flashers such as fireflies and glow-worms, and dancers such as cranes and grebes. The male great crested newt wafts female hormones at his intended in order to arouse her, while the male Fowler's toad makes use of the laws of acoustics and sits in the cooler parts of the

NUPTIAL GIFT *The male robber fly presents the female with a cottonstainer bug to eat in order to distract her during mating.*

first group were noticeably smaller and less well-nourished than those of the second, suggesting that a nutritious substance, in addition to the sperm, had been transferred from the male to the female. The identity of the substance has still to be discovered.

The courting male cricket makes a more obvious gift to his partner. Along with the transfer of a sperm capsule, known as a spermatophore, the male also supplies the female with a gelatinous blob, known as a spermatophylax. The purpose of the gift is twofold: firstly, it provides the female with additional food that can be used to make more eggs, and secondly, while she is eating it she is less likely to mate with another male, ensuring that the first male's sperm has a chance to work.

FEMALE CHOICE

In some species it is the female that starts things going. Female moths release a scent that can be picked up by a male moth's feathery antennae from many miles away. He can detect just one molecule and flies upwind until he finds the female in question. She releases the chemical in a pulsed fashion rather than in a single stream. Undistracted by the pulsed emissions of other females, the male following her vapour trail homes in exclusively on her.

Some moths, such as the corn earworm moth, must coordinate courtship and mating with the availability of food for their offspring. The corn earworm caterpillar feeds, as its name suggests, on the ears of corn,

has been known to bite off her partner's head and devour his body while still in the act of procreation. Male spiders and some species of fly avoid a similar fate by appeasing their partners with a nuptial gift.

Male robber flies scour the air for small prey, usually other flies, capture it, and present it to their intended. The female sucks the fly dry while the male goes about his reproductive business. Similarly, male empid flies, such as *Hilaria*, capture prey for females and gift-wrap their offering with a few strands of silk. *Hilaria* produces the silk itself.

The gifts offered by crickets and bowl-and-doily spiders help nourish the female's

future offspring. Scientists noticed that bowl-and-doily spiders, named for the complex webs they produce, complete sperm transfer in just 15 minutes, yet two spiders will remain embraced for up to an hour. In a laboratory experiment, pairs of mating spiders were divided into two groups. The pairs in the first group were separated after just 15 minutes, while those in the other group were allowed to let nature run its course. The offspring of the

CONJUGAL JELLY *A female grasshopper is preoccupied with her mate's gift of food and is less likely to mate again before his sperm fertilises her eggs.*

FOLLOW THE SCENT TRAIL
The large antennae on the head of the male emperor moth are able to detect a single molecule of a female's scent.

and so the adult moths must synchronise reproduction with the most advantageous phase in the plant's growing cycle. They do this in response to chemicals produced by the plant, including the hormone ethylene, which induces fruit-ripening. The plant hormone triggers the production of hormones in the female, which are then released to attract a male. After mating, she deposits her eggs on the ripening corn.

The use of scent is a widespread means of attracting a mate – female garter snakes, for example, release a scent that will have them smothered by a writhing mass of up to 150 males in seconds. Male snakes also mimic the females by producing the same sex pheromone. One doing this also becomes immediately surrounded by males, but during the ensuing melee the 'she-male' slips surreptitiously away and has the nearest female all, or almost all, to himself.

Smell, though, is not the only channel of communication available to female animals to advertise their readiness to mate. Female elephants make very low-frequency calls, below the range of human hearing, and summon every bull elephant in musth (a state of sexual excitement) for many miles around. The males must reach her rapidly for she will only be receptive for five minutes once every three or four years.

In the majority of species, it is the female

RED TALK

Some species of nymphalid and pierid butterflies are able to see the broadest visible spectrum of any known animal. Their compound eyes are sensitive to light extending from the ultraviolet to the far red. It seems that the bright orange, red and yellow colours, once thought to be warning colours solely for deterring predators, are involved in communication between individuals of the same species. Male and female butterflies talk to each other in the far red end of the light spectrum.

that invests most time and energy in rearing the young. She also chooses the partner, selecting the best singer or dancer or best-dressed performer for his fitness to provide her offspring with the best possible genes.

'FEMME-FATALE' FIREFLIES

The sneakiest female must be the 'femme-fatale' firefly. Like all fireflies, she flashes a coded signal to circling male fireflies, and they home in on her glow in order to mate. Normally females flash the code of their own species in response to a male's flash, but this particular firefly cheats. Male fireflies approach a bush containing females and give the species signal. *Photinus pyralis* males, for example, flit an undulating course just 20 in (51 cm) above the ground, drop down every 5.8 seconds precisely, and emit a half-second flash. Any female within 7 ft (2 m) waits for 2 minutes and then flashes back. By matching flashes, fireflies find a mate of their own species.

The 'femme-fatale' females of the genus *Photuris*, however, give the *Photinus* reply when they spot a *Photinus* male's flash. The male flies down expecting to mate and, to his immense surprise, is killed and eaten. He has something that the *Photuris* female needs. When seeking a place to deposit their eggs, female fireflies are vulnerable to wolf spiders and ants. Both the male and female *Photinus* have poisonous steroids in their blood that discourage predators from eating them. *Photuris* females have not developed the ability to synthesise them, however. Instead, they capture *Photinus* males and acquire their protection from them.

LIVING LIGHTS *A tree is lit up by hundreds of fireflies, which are not flies but beetles with a luminous tip on the abdomen.*

In a few species, such as the crested auklet, both sexes devote time to guarding and feeding the next generation.

Both male and female crested auklets, which breed on the Alaskan coast, have crests. At one time it was thought that the female crests were redundant features, only present because the genes that created them were inextricably linked up with those responsible for the male's crest. Recent research shows this not to be the case. Males and females search for members of the opposite sex with the best crests. Both sexes use the crests as an indicator of an individual's quality and as a means of choosing the best available partner.

COMMITTED COMPANIONS

Some partners remain together for life. Swans and geese are well-known examples, but clown shrimps, California mice, Djungarian hamsters and titi monkeys are monogamous, too. Monogamy can be advantageous for some. During the breeding season time is not lost in searching for a mate and a breeding site, and two adults rather than one mean that offspring are well looked after. In the brief Arctic summer, for instance, it pays swans, geese and terns to have a partner available on arrival so that vital breeding and feeding time is not wasted. But all is not as it seems.

The Djungarian or dwarf hamster, unlike its golden (pet) hamster relatives, which are fiercely solitary, live in pairs; the male assists the female in looking after the new pups. During the first few days of life, baby dwarf hamsters have little control over their body temperature and must be cuddled constantly by a parent, often the father. But here the blissful picture ends, for

the family life of the dwarf hamster is worthy of a television soap opera.

Each female hamster shares separate burrows with at least two other males, and each male shares a burrow with at least two different females. Each pair, however, has a single burrow that only they occupy. They might both have yet another burrow that each inhabits alone. When she is ready to mate, a female will visit the burrows of all her males, although a pair will share a burrow on the day of sleep that precedes the evening she is ready to mate. As if this were not complicated enough, she is also visited by other males, which travel over a mile to be with her during the evening. The males fight with each other, but rarely supplant the resident male.

Unlike the dwarf hamsters, some animals appear to be faithful partners, but their attentions are purely selfish. A blackbird stays close to his mate to make sure another male does not copulate with her when she is fertile.

SHRIMP COUPLE *The male freshwater shrimp attaches himself to his chosen female, and then waits until she moults before he can mate.*

Sparrows and ospreys copulate frequently in an attempt to ensure that their sperm overwhelms that of any rival who might have sneaked in while the male was away.

Some creatures must stay close to their betrothed in case they miss the opportunity to breed. The $1/2$ in (1.3 cm) long, curved male *Gammarus* freshwater shrimp, for example, can only copulate with a female when she has moulted, which occurs every three weeks or so. All the females in a population do not moult at the same time, so the male can either swim among the population of females in the hope that he chances upon a moulting female, or seize a suitable female and lock on to her until she is ready to moult.

EGGS AND EMBRYOS

Offspring can be wrapped in a protective, nourishing egg or retained inside the mother's body. The former tend to be produced in large numbers, while the parents of the latter can afford to put all their energy into fewer babies.

Animals adopt one of two main strategies in bringing the new brood into the world: they produce either a large number of offspring that, to all intents and purposes, are left to fend for themselves, or a few well-protected and well-nourished eggs or embryos.

The giant clam opts for the first strategy. At breeding time it ejects 500 million eggs, the most eggs produced by any living organism on earth. They drift with the ocean currents, hatch into free-floating larvae and provide food for small marine

GOLDEN CAVIAR *Pairs of black embryonic eyes peep from the eggs of the lumpsucker fish as they float in a large cluster.*

creatures. Of the millions that start out, only a few survive to settle and grow into new giant clams.

The million eggs produced by each cod, herring or pollack, however, are not totally helpless. Pollack eggs are able to alter the depth at which they float by changing their buoyancy. In this way, they find the best conditions in which to develop. Four days after release, and in response to light, the eggs of the walleye pollack, which is found in the Pacific off Alaska, become more dense and sink in the water column. They continue to fall until they reach a depth of about 500 ft (150 m). Here they are relatively

safe from surface predators and protected from dangerous ultraviolet light, which can damage DNA, the vital blueprint of life in each cell. If the eggs should be swept towards the surface in an upwelling current, the change in buoyancy, triggered by the light at the surface, ensures that they sink back down to the safety of the ocean depths.

ABANDONED NESTS

Some mothers abandon their eggs, but not until they have ensured that the eggs are safe. Mother sea turtles haul themselves laboriously from the sea, usually returning to the beach where they hatched out many years before. They crawl to the parts of the beach where the sand is just the right dampness and temperature and dig a large hole, into which they deposit their 100 or so eggs. After covering the nest they return to the sea, leaving the eggs and later the hatchlings to fend for themselves.

Long-necked freshwater turtle mothers in the Northern Territory of Australia break all the rules and deposit their eggs under water. This is unusual behaviour, for reptile eggs need oxygen from the air to survive. The long-necked turtle, however, lives in a region where the weather can swing between monsoon conditions and no rain at all, leaving a turtle mother to cope with either floods or parched ground. Her solution is to bury her eggs to a depth of 6-8 in (15-20 cm) in flooded ground at the edge of a lagoon. Because of the lack of oxygen here, their development is arrested until the dry season arrives and the water evaporates. The eggs develop under the hard mud and are ready to hatch six months

later when the rains return and the plain is flooded once more.

Puddle or tungara frogs in the rainforests of South and Central America also have water problems. Despite the deluge that falls on them daily, there is very little standing water, such as lakes and ponds, in which to deposit eggs. The only bodies of water available are temporary rain pools,

which are also oases for predators. In order to give their offspring the best start in life, tungara frogs provide a protective foam nest. They whisk up the egg jelly with their hind legs into a dense foam and deposit the eggs in the centre. Here they are safe from predators and from desiccation. In one species, the savannah frog, the tadpoles continue to benefit from the foam, where

they remain concealed for several days after hatching. If the foam begins to dry out, they add more by producing mucus and beating it with their tales.

The little dragon sculpin, a fish of the North Pacific, has found one of the safest places for her eggs. She lays them inside a living sponge of the genus *Mycale*, boring into the soft exoskeleton and depositing clumps of three to fifteen eggs in the canals that permeate the sponge. The eggs, which may take up to eight months to hatch, are bathed in oxygen-rich sea water and are protected by the fungicides and antibacterial chemicals produced by the sponge.

Stick insects on land also bury their eggs away from the attentions of tiny parasitic wasps, but instead of doing the work themselves, they enlist the help of ants,

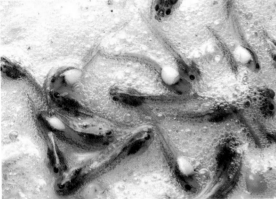

BUBBLE BATH *The tadpoles of the South African foam-nest tree frog (left) are hidden in a protective mass of bubbles. Below: A pair of tungara frogs beat up a foam nest with their back legs, and deposit and fertilise their eggs in the froth.*

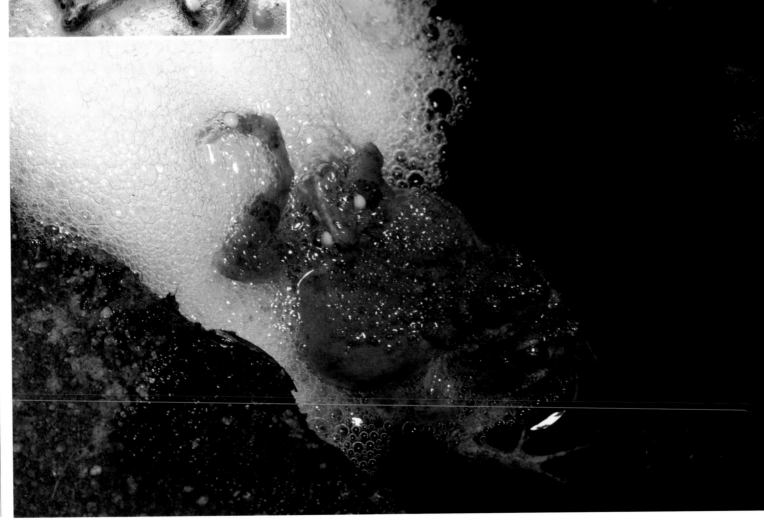

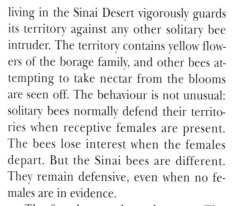

Seed Bearer Right: Some seeds have a fatty nodule to encourage ants to carry them back to the nest, where conditions are right for germination. Above: The stick insect egg has a white knob that mimics the nodule on seeds in the hope that ants will carry it back to their nest.

which take their cue from plants. Many species of ant help plants to distribute their seeds, but they require a reward. The plant has evolved small structures on its seeds, known as elaiosomes, which contain fatty substances favoured by ants. Seeds providing such tempting nourishment are collected by the ants and taken back to the nest, where the ants eat the fatty substance. Inside the nest, the seed germinates, and the plant has successfully propagated itself. Some species of stick insect, including the laboratory stick insect often kept as a pet, have adopted the same approach.

The stick insect's eggs have a small white or yellow projection, known as the capitulum, on the top of the tiny hatch, or operculum, where the nymph emerges. The ants grab the eggs and carry them back to the nest, just as they do with seeds. Eggs missed by the ants and left on the surface of the leaf litter have a 20 per cent greater chance of being attacked by a parasitic wasp than those taken back to the ants' nest.

GUARD DUTY

It is well known that in many species of bird the parents share nest-guarding duties, but a species of solitary bee is just as diligent. The bee – yet to be given a scientific name –

living in the Sinai Desert vigorously guards its territory against any other solitary bee intruder. The territory contains yellow flowers of the borage family, and other bees attempting to take nectar from the blooms are seen off. The behaviour is not unusual: solitary bees normally defend their territories when receptive females are present. The bees lose interest when the females depart. But the Sinai bees are different. They remain defensive, even when no females are in evidence.

The females are there, however. The male and female are simply patrolling the territory in shifts. The female takes the early morning and late afternoon shifts, while the male takes the long midday shift. The male guards the territory while the plants are building up supplies of nectar, and the female gathers nectar and pollen at the optimum period during the afternoon. She mixes the pollen and nectar into a food ball and places an egg on it. The food ball and egg are deposited in an underground nest in the male's territory. The male continues to guard the territory, enabling the female to complete her task, until the breeding season is over.

Some parents opt out entirely and dupe

other parents into raising their young. Birds are particularly good at this, and the arch villain of the avian world is probably the European cuckoo.

Parent cuckoos do not form a lasting relationship, but the male nonetheless plays an important role in the process of deception. After mating, the female identifies several suitable host nests. In Britain these are usually the nests of meadow pipits on moorland, reed warblers in marshland, dunnocks in farmland and pied wagtails on open habitats. A cuckoo and its offspring will specialise in one species, and ornithologists now refer to them as pipit-cuckoos or pied wagtail-cuckoos, depending on their preferred host.

The deception starts in the afternoon. Timing is critical. The male cuckoo creates a diversionary rumpus guaranteed to bring an enraged resident away from one of the targeted nests. Meanwhile the female cuckoo, who has been hiding nearby, flies to the nest. She picks up a host's egg, deposits one of her own, flies quickly away and consumes the host's egg. The entire procedure takes no more than 15 seconds, and this is repeated at several nests during the day.

Nest parasitism and egg dumping, in

PREGNANT MALE

In some species it is the fathers that are left holding the baby. The male seahorse, for instance, undergoes a complete pregnancy. After a long courtship the female seahorse deposits her eggs in a pouch on the male's abdomen and then leaves him to look after their offspring. The male fertilises the eggs, which then attach to the pouch wall. The wall is well supplied with blood vessels, which provide the eggs containing the developing embryos with oxygen. The eggs hatch, and the larval seahorses are retained in the pouch. Here they are nourished by a fluid secreted by the male. After a period of up to 50 days, the young are forcibly ejected from the pouch. The male seahorse might pump out up to 14 000 youngsters, each only $1/4$ - $3/8$ in (6-10 mm) long, the entire birth process taking up to 48 hours.

SHARK EGGS AND EMBRYOS

Sharks share their breeding strategy with whales and human beings. They grow to a large size, and invest considerable time and energy in raising resilient offspring. All sharks fertilise their eggs internally and produce relatively few offspring at any one time. And they have resorted to many different ways of giving their pups the best start in life.

Common dogfish and nursehounds deposit eggs, each developing embryo protected in a tough, leathery egg-case – the familiar 'mermaid's purse' – with tendrils to anchor it in place. The Port Jackson shark has a more ingenious anchoring mechanism. Each egg-case – 6 in (15 cm) in length – has two spiral flanges wrapped around the outside. The mother shark moves into shallow water and wedges the egg into crevices between rocks, the screw-like flanges taking a firm hold in the crack.

Many shark mothers play even safer. They retain their embryos inside the body until the embryos

PROTECTIVE TWIST *The female Port Jackson shark pushes her egg-case, which is flanged like a wood screw, into a crack in the rocks.*

are sufficiently well developed to fend for themselves – a process known as viviparity.

The offspring of the piked or spiny dogfish begin their development enclosed in an amber-coloured 'candle'. Five fertilised eggs are stacked inside a long, thin, candle-shaped membrane that is retained within the mother's body. After about six months the candle ruptures or dissolves and the embryos continue to grow in the uterus for a further 14 months, each one living on its own yolk sac. After a 22 month gestation period, ten pups, 10 in (25 cm) in length, emerge.

Hammerhead sharks take things a stage further. The embryos are retained inside the mother and their first source of nutrition is the yolk sac. After about four months, however, the sac begins to grow branches and fastens to the uterine wall in the manner of a 'placenta'. Nutrients and oxygen are diffused from the maternal blood supply via the placenta to the embryo. The young are born with umbilical scars, which disappear after a few weeks.

The most bizarre form of viviparity is shown by the ragged-toothed sharks, such as sand tiger and nurse sharks.

A bunch of fertilised eggs takes its place in each of the shark's pair of uteri, and the embryos develop in the usual way, but they quickly use up the food in their yolk sacs. Then a dominant embryo in each uterus devours its womb-mates. The mother continues to ovulate and the surviving pair of embryos feast on the steady

MERMAID'S PURSE *The egg-case of the common dogfish, containing a visible embryo, is anchored to rocks and seaweed by tendrils at each corner.*

stream of unfertilised eggs. Eventually the two intra-uterine cannibals grow to a sufficient size to be born.

HOODWINKED *A rufous-collared sparrow parent is unwittingly bringing up the chicks of an entirely different species, the parasitic cowbird.*

which parents literally dump their eggs in the nests of others, are not uncommon in birds. About 1 per cent of all bird species do it, including cowbirds, honeyguides, buffalo weavers, bearded tits, whydahs, lapwings, swallows, mergansers, ducks and geese, but other creatures, such as garden snails, are egg-dumpers too.

Usually a garden snail will dig laboriously for two days, deposit its clutch of eggs and then cover the nest. Given the opportunity, however, snails take the easy way out. They dig into the nests of other snails and deposit their eggs there. This alternative strategy takes less than a day and saves the snail a considerable amount of energy, which it channels into producing more eggs. Egg-dumpers consistently produce more eggs each season than nest diggers.

Fish are also egg-dumpers. Certain 'lazy' species take advantage of other, more

SNAIL'S BASE *Common European garden snails deposit their eggs in the nests of other snails, saving themselves time and energy.*

conscientious parents, and the story of the perch and the minnow is a case in point. The male Japanese freshwater perch is a good parent. He sees off rival perch and establishes a territory about 3 ft (1 m) in diameter. He clears algae and other debris from several reed stems and then invites several females to deposit their eggs there. He fans them with his fins and defends them rigorously against any nest predators. The females depart, leaving him to care for the brood. Then, all of a sudden, a band of 20 marauding minnows descend on his patch and in a frantic orgy of mass spawning deposit their eggs alongside his own. As the desperate resident attempts to evict the intruders, a species of chub sneaks into his territory and eats the perch's eggs while his

HATCHING TIME *More garden snails hatch from eggs deposited by nest parasites than from the parent snails that dug the nest site.*

attention is elsewhere. The perch then ends up guarding mainly the eggs of the minnows and a few of his own.

An even more bizarre story of dumping involves fry rather than eggs. The species involved is the convict cichlid, named for its distinctive body stripes. Convict cichlid parents are monogamous. They set up territory in a shallow stream and excavate a small cave in which they deposit and fertilise their eggs. When the fry hatch out they are guarded by both parents. They are escorted outside to feed during the day and then ushered back into the cave at night.

Quite often, however, the number of fry returning to the cave in the evening is greater than the number that left in the morning. During the day a single parent, having lost a partner, usually the male, through desertion or predation, has put out its brood for adoption by the resident pair. A single parent is less able to defend its brood, and adoption is one way in which it can offer some of them help to survive.

Resident cichlids, however, are choosy about which broods they will accept for adoption. The larger brood is welcomed in principle because increased numbers

means that the resident fish are less likely to lose offspring. The preference, however, is for fry that are smaller than their own. Predators have a preference for smaller, slow-swimming fry, leaving the larger individuals alone. By adopting another cichlid's brood of smaller fry, the resident fish minimise attacks on their own offspring.

THE END

For some parents the effort of procreation is all too much. After it is all over, they die. Terminal spawning, as it is known, is the ultimate goal of the squid. The deep-sea squid *Moroteuthis*, for example, lives at depths between 2428 ft (740 m) and 4757 ft (1450 m) off the coast of New Zealand. The male has huge testes that fill much of the body cavity and a penis that is longer than the main body. It fills with spermatophores, ready to transfer to a female. The mature female's body is also dominated by her reproductive organs. She is much larger than the male and her ovaries alone weigh as much as a full-grown male.

When the sperm and eggs are shed, neither male nor female squid feeds again. A gland situated near the eye is activated by hormones released from the testes and ovaries, and the squids are henceforth unable to synthesise proteins. The result is a dramatic breakdown in muscle tissue and the ultimate death of both parent squids.

The reason for terminal spawning is not clear, although a single session during which large numbers of young are produced may give the young a better chance of surviving than several sessions producing less spawn, as predators are swamped by the deluge of squid offspring and become quickly sated. The mother, however, in marshalling all her reserves of energy in order to provide her offspring with the best start in life, has paid the ultimate price: she has sacrificed her own life.

EXPANDING NURSERY *A convict cichlid may find its brood increases as other parents dump their young with them. Overleaf: Salmon, in bright red courtship colours, head upstream to spawn in the headwaters of the rivers they left four years previously.*

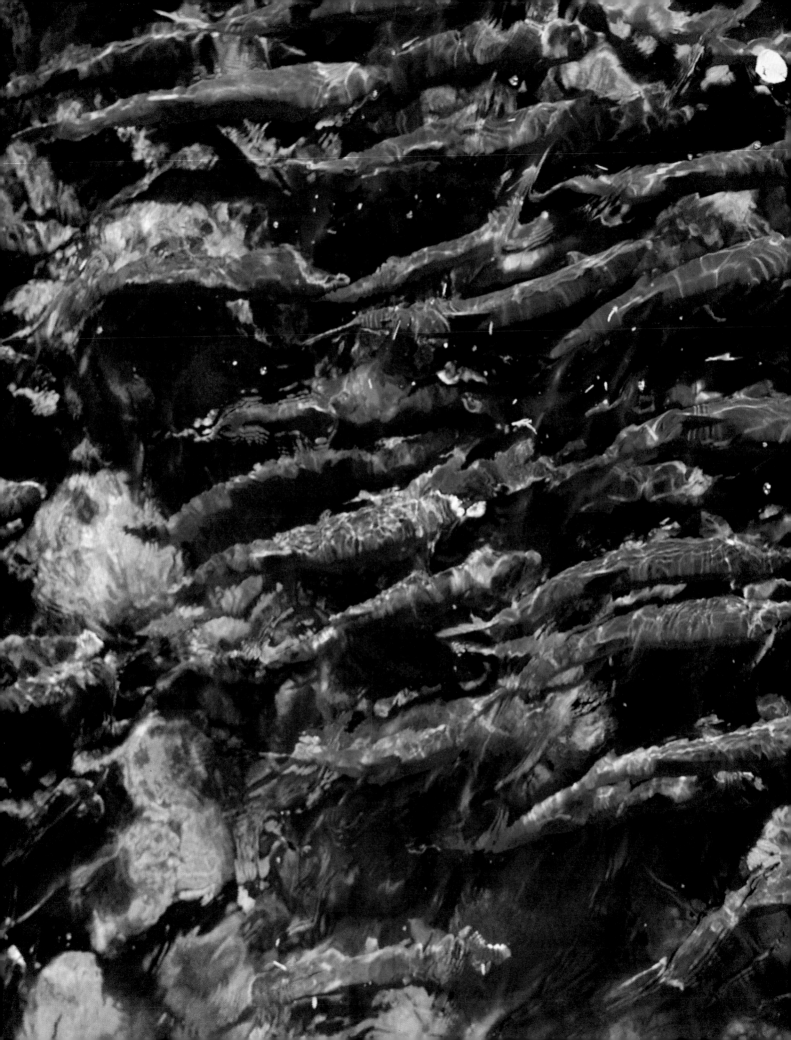